REVIEW OF THE ARMY NON-STOCKPILE
CHEMICAL MATERIEL DISPOSAL PROGRAM

Disposal of Neutralent Wastes

COMMITTEE ON REVIEW AND EVALUATION OF THE ARMY
NON-STOCKPILE CHEMICAL MATERIEL DISPOSAL PROGRAM

BOARD ON ARMY SCIENCE AND TECHNOLOGY

DIVISION ON ENGINEERING AND PHYSICAL SCIENCES

NATIONAL RESEARCH COUNCIL

NATIONAL ACADEMY PRESS
Washington, D.C.

National Academy Press • 2101 Constitution Avenue, N.W. • Washington, D.C. 20418

NOTICE: The project that is the subject of this report was approved by the Governing Board of the National Research Council, whose members are drawn from the councils of the National Academy of Sciences, the National Academy of Engineering, and the Institute of Medicine. The members of the committee responsible for the report were chosen for their special competencies and with regard for appropriate balance.

This is a report of work supported by Contract DAAG55-98-C-0046 between the U.S. Army and the National Academy of Sciences. Any opinions, findings, conclusions, or recommendations expressed in this publication are those of the author(s) and do not necessarily reflect the view of the organizations or agencies that provided support for the project.

International Standard Book Number: 0-309-07287-5
Library of Congress Catalog Card Number: 2001087475

Limited copies are available from:

Board on Army Science and Technology
National Research Council
2101 Constitution Avenue, N.W.
Washington, DC 20418
(202) 334-3118

Additional copies are available for sale from:

National Academy Press
Box 285
2101 Constitution Ave., N.W.
Washington, DC 20055
(800) 624-6242 or (202) 334-3313
(in the Washington Metropolitan area)
http://www.nap.edu

Copyright 2001 by the National Academy of Sciences. All rights reserved.
Printed in the United States of America.

THE NATIONAL ACADEMIES

National Academy of Sciences
National Academy of Engineering
Institute of Medicine
National Research Council

The **National Academy of Sciences** is a private, nonprofit, self-perpetuating society of distinguished scholars engaged in scientific and engineering research, dedicated to the furtherance of science and technology and to their use for the general welfare. Upon the authority of the charter granted to it by the Congress in 1863, the Academy has a mandate that requires it to advise the federal government on scientific and technical matters. Dr. Bruce M. Alberts is president of the National Academy of Sciences.

The **National Academy of Engineering** was established in 1964, under the charter of the National Academy of Sciences, as a parallel organization of outstanding engineers. It is autonomous in its administration and in the selection of its members, sharing with the National Academy of Sciences the responsibility for advising the federal government. The National Academy of Engineering also sponsors engineering programs aimed at meeting national needs, encourages education and research, and recognizes the superior achievements of engineers. Dr. William A. Wulf is president of the National Academy of Engineering.

The **Institute of Medicine** was established in 1970 by the National Academy of Sciences to secure the services of eminent members of appropriate professions in the examination of policy matters pertaining to the health of the public. The Institute acts under the responsibility given to the National Academy of Sciences by its congressional charter to be an adviser to the federal government and, upon its own initiative, to identify issues of medical care, research, and education. Dr. Kenneth I. Shine is president of the Institute of Medicine.

The **National Research Council** was organized by the National Academy of Sciences in 1916 to associate the broad community of science and technology with the Academy's purposes of furthering knowledge and advising the federal government. Functioning in accordance with general policies determined by the Academy, the Council has become the principal operating agency of both the National Academy of Sciences and the National Academy of Engineering in providing services to the government, the public, and the scientific and engineering communities. The Council is administered jointly by both Academies and the Institute of Medicine. Dr. Bruce M. Alberts and Dr. William A. Wulf are chairman and vice chairman, respectively, of the National Research Council.

COMMITTEE ON REVIEW AND EVALUATION OF THE ARMY NON-STOCKPILE CHEMICAL MATERIEL DISPOSAL PROGRAM

JOHN B. CARBERRY, chair, E.I. duPont de Nemours and Company, Wilmington, Delaware
JOHN C. ALLEN, Battelle Memorial Institute, Washington, D.C.
RICHARD J. AYEN, Waste Management, Inc. (retired), Wakefield, Rhode Island
ROBERT A. BEAUDET, University of Southern California, Los Angeles
LISA M. BENDIXEN, Arthur D. Little, Inc., Cambridge, Massachusetts
JOAN B. BERKOWITZ, Farkas Berkowitz and Company, Washington, D.C.
JUDITH A. BRADBURY, Pacific Northwest National Laboratory, Washington, D.C.
MARTIN C. EDELSON, Ames Laboratory, Ames, Iowa
SIDNEY J. GREEN, TerraTek, Inc., Salt Lake City, Utah
PAUL F. KAVANAUGH, consultant, Fairfax, Virginia
DOUGLAS M. MEDVILLE, MITRE (retired), Reston, Virginia
WINIFRED G. PALMER, Henry M. Jackson Foundation for the Advancement of Military Medicine, Bethesda, Maryland
JAMES P. PASTORICK, GEOPHEX UXO, Alexandria, Virginia
WILLIAM J. WALSH, Pepper Hamilton LLP, Washington, D.C.
RONALD L. WOODFIN, Sandia National Laboratories (retired), Albuquerque, New Mexico

Staff

STERLING J. RIDEOUT, JR., study director
DELPHINE D. GLAZE, administrative assistant
GREG EYRING, consultant
CAROL R. ARENBERG, editor, Division on Engineering and Physical Sciences

BOARD ON ARMY SCIENCE AND TECHNOLOGY

WILLIAM H. FORSTER, chair, Northrop Grumman Corporation, Baltimore, Maryland
RICHARD A. CONWAY, Union Carbide Corporation (retired), Charleston, West Virginia
GILBERT F. DECKER, Walt Disney Imagineering, Glendale, California
PATRICK F. FLYNN, Cummins Engine Company, Inc., Columbus, Indiana
EDWARD J. HAUG, University of Iowa, Iowa City
ROBERT J. HEASTON, Guidance and Control Information Analysis Center (retired), Naperville, Illinois
GERALD J. IAFRATE, University of Notre Dame, Notre Dame, Indiana
DONALD R. KEITH, Cypress International, Alexandria, Virginia
KATHRYN V. LOGAN, U.S. Army Corps of Engineers, Vicksburg, Mississippi
JOHN E. MILLER, Oracle Corporation, Reston, Virginia
JOHN H. MOXLEY, Korn/Ferry International, Los Angeles, California
STEWART D. PERSONICK, Drexel University, Philadelphia, Pennsylvania
MILLARD F. ROSE, NASA Marshall Space Flight Center, Huntsville, Alabama
GEORGE T. SINGLEY, III, Hicks and Associates, Inc., McLean, Virginia
CLARENCE G. THORNTON, Army Research Laboratories (retired), Colts Neck, New Jersey
JOHN D. VENABLES, Venables and Associates, Towson, Maryland
JOSEPH J. VERVIER, ENSCO, Inc., Melbourne, Florida
ALLEN C. WARD, Ward Synthesis, Inc., Ann Arbor, Michigan

Staff

BRUCE A. BRAUN, director
MICHAEL A. CLARKE, associate director
WILLIAM E. CAMPBELL, administrative coordinator
CHRIS JONES, financial associate
REBECCA M. LUCCHESE, senior project assistant
DEANNA P. SPARGER, senior project assistant

Preface

The Committee on Review and Evaluation of the Army Non-Stockpile Chemical Materiel Disposal Program was appointed by the National Research Council (NRC) to conduct studies on technical aspects of the U.S. Army Non-Stockpile Chemical Materiel Disposal Program. During its first year, the committee evaluated the Army's plans to dispose of chemical agent identification sets (CAIS)—test kits used for training soldiers. During this second year, the committee has evaluated nonincineration technologies that could be used for the treatment of wastes from the neutralization of nonstockpile materiel.

During its initial meetings, the committee was given a number of briefings and held subsequent deliberations. The committee is grateful to the many individuals, particularly Lt. Col. Christopher Ross, Project Manager for Non-Stockpile Chemical Materiel, his staff, and his predecessor, Col. Edmund W. ("Ned") Libby, who provided technical information and insights during these briefings. This information provided a sound foundation for the committee's deliberations.

This study was conducted under the auspices of the NRC's Board on Army Science and Technology. The committee acknowledges the support of the director, Bruce A. Braun, his staff, committee members, the study director, support staff, and the publication staff who all worked diligently on a demanding schedule to produce this report.

John B. Carberry, chair
Committee on Review and Evaluation of the
Non-Stockpile Chemical Materiel Disposal Program

Acknowledgments

This report has been reviewed by individuals chosen for their diverse perspectives and technical expertise, in accordance with procedures approved by the NRC's Report Review Committee. The purpose of this independent review is to provide candid and critical comments to assist the authors and the NRC in making the published report as sound as possible and to ensure that the report meets institutional standards for objectivity, evidence, and responsiveness to the study charge. The content of the review comments and the draft manuscript remain confidential to protect the integrity of the deliberative process. We wish to thank the following individuals for their participation in the review of this report:

James R. Fair, University of Texas
Richard S. Magee, New Jersey Institute of Technology
John L. Margrave, Rice University
Walter G. May, University of Illinois
Alvin H. Mushkatel, Arizona State University

George W. Parshall, E.I. duPont de Nemours and Company (retired)
Michael J. Ryan, NFT, Inc.
R. Peter Stickles, Arthur D. Little, Inc.
William Tumas, Los Alamos National Laboratory
Leo Weitzman, LVW Associates, Inc.

Although the reviewers listed above provided many constructive comments and suggestions, they were not asked to endorse the conclusions or recommendations nor did they see the final draft of the report before its release. The review of this report was overseen by Robert Connick, appointed by the NRC Report Review Committee, who was responsible for making certain that an independent examination of this report was carried out in accordance with institutional procedures and that all review comments were carefully considered. Responsibility for the final content of this report rests entirely with the authoring committee and the institution.

Contents

EXECUTIVE SUMMARY ... 1

1 OVERVIEW ... 6
    Chemical Stockpile Disposal Program, 6
    Non-Stockpile Chemical Materiel Disposal Program, 7
    Applicability of ACWA Technologies, 10
    Role of the National Research Council, 10
    Scope of This Study, 10
    Structure of This Report, 11

2 WASTE STREAMS FROM TRANSPORTABLE TREATMENT SYSTEMS ... 12
    Chemical Agent Identification Sets, 12
    Rapid Response System, 12
    Munitions Management Device, 13
    Toxicity of Neutralents, 14
    Federal and State Requirements, 15

3 CRITERIA FOR EVALUATING TECHNOLOGIES ... 20
    Temperature Classifications, 20
    Pressure Classifications, 20
    Selection Criteria, 20
    Top Priority Criteria, 21
    Important Criteria, 22

4 DESCRIPTIONS AND EVALUATIONS OF TECHNOLOGIES ... 23
    Chemical Oxidation, 23
    Electrochemical Oxidation, 26
    Biodegradation, 30
    Solvated-Electron Technology, 32
    Wet-Air/O_2 Oxidation, 34
    Supercritical Water Oxidation, 37
    Gas-Phase Chemical Reduction, 39
    Plasma-Arc Technology, 41
    Overall Rankings, 43

5 PUBLIC ACCEPTANCE AND REGULATORY CONSIDERATIONS   47
    Public Acceptance, 47
    Regulatory Stakeholders, 49

6 FINDINGS AND RECOMMENDATIONS   52
    Technical Issues, 52
    Regulatory Issues and Public Involvement, 53

REFERENCES   54

APPENDIXES
    A   Biographical Sketches of Committee Members, 59
    B   Committee Meetings and Other Activities, 62

Figures and Tables

FIGURES

1-1 Flow chart for the disposal of nonstockpile CWM in transportable Systems, 9

4-1 Comparative operating temperatures and pressures, 44

TABLES

ES-1 Transportable Treatment Systems and Neutralent Waste Streams Considered in This Study, 2

1-1 Transportable Treatment Systems for Nonstockpile Chemical Materiel, 8

2-1 CAIS Chemical Agents and Treatment Processes in the Rapid Response System, 13
2-2 Composition of Neutralent Waste Streams from the Rapid Response System, 14
2-3 Reagents Used to Neutralize Chemical Agents in the MMD, 15
2-4 Composition of Sarin (GB) Neutralent Wastes from Bench-Scale Tests of the MMD, 16
2-5 Composition of Mustard (HD) Neutralent Wastes from Bench-Scale Tests of the MMD, 17
2-6 Composition of VX Neutralent Wastes from Bench-Scale Tests of the MMD, 18
2-7 Composition of Phosgene Neutralent Wastes from Bench-Scale Tests of the MMD, 18
2-8 Toxicity of Components of the O/SSs Used in the RRS and MMD, 19

3-1 Technologies Selected for Evaluation, 21

4-1a Chemical Oxidation: Top Priority Criteria, 24
4-1b Chemical Oxidation: Important Criteria, 25
4-2a Electrochemical Oxidation Ag(II): Top Priority Criteria, 27
4-2b Electrochemical Oxidation Ag(II): Important Criteria, 28
4-3a Electrochemical Oxidation Ce(IV): Top Priority Criteria, 29
4-3b Electrochemical Oxidation Ce(IV): Important Criteria, 30
4-4a Biodegradation: Top Priority Criteria, 31
4-4b Biodegradation: Important Criteria, 32
4-5a Solvated-Electron Technology: Top Priority Criteria, 33
4-5b Solvated-Electron Technology: Important Criteria, 34
4-6a Wet-Air/O_2 Oxidation: Top Priority Criteria, 35
4-6b Wet-Air/O_2 Oxidation: Important Criteria, 36
4-7a Supercritical Water Oxidation: Top Priority Criteria, 38
4-7b Supercritical Water Oxidation: Important Criteria, 39

4-8a Gas-Phase Chemical Reduction: Top Priority Criteria, 40
4-8b Gas-Phase Chemical Reduction: Important Criteria, 41
4-9a Plasma-Arc Technology: Top Priority Criteria, 42
4-9b Plasma-Arc Technology: Important Criteria, 43

Acronyms and Abbreviations

ACWA	Assembled Chemical Weapons Assessment (Program)	NSCWCC	Non-Stockpile Chemical Weapons Citizens Coalition
ATAP	Alternative Technology Approaches Program	O/SS	oxidant/solvent system
CAIS	chemical agent identification sets	PCB	polychlorinated biphenyl
CWC	Chemical Weapons Convention	PINS	portable isotopic neutron spectroscopy
CWM	chemical warfare materiel	PMCD	Program Manager for Chemical Demilitarization
CVA	chlorovinylarsonic acid		
DCDMH	dichloro-dimethylhydantoin	POTW	publicly owned treatment works
DOT	U.S. Department of Transportation	ppb	parts per billion
		ppm	parts per million
EDS	explosive destruction system	psia	pounds per square inch absolute
EPA	Environmental Protection Agency		
		RCRA	Resource Conservation and Recovery Act
FOTW	federally owned treatment works	RRS	rapid response system
GB	sarin (nerve agent)	SCWO	supercritical water oxidation
GPCR	gas-phase chemical reduction	SET	solvated-electron technology
H	sulfur mustard	TCLP	toxic characteristic leaching procedure
HD	sulfur mustard (distilled)	TSDF	treatment, storage, and disposal facility
HN-1, H-3	nitrogen mustard		
		UV	ultraviolet
MEA	monoethanolamine		
MMD	munitions management device	VOC	volatile organic compound
		VX	a nerve agent
NATO	North Atlantic Treaty Organization		
NSCMP	Non-Stockpile Chemical Materiel Program	WAO	wet-air/O_2 oxidation

Executive Summary

Chemical warfare materiel (CWM) is a collection of diverse items that were used during 60 years of efforts by the United States to develop a capability for conducting chemical warfare. Nonstockpile CWM, which is not included in the current U.S. inventory of chemical munitions, includes buried materiel, recovered materiel, binary chemical weapons, former production facilities, and miscellaneous materiel. CWM that was buried in pits on former military sites is now being dug up as the land is being developed for other purposes. Other CWM is on or near the surface at former test and firing ranges. According to the Chemical Weapons Convention (CWC), which was ratified by the United States in April 1997, nonstockpile CWM items in storage at the time of ratification must be destroyed by 2007.[1]

The U.S. Army is the designated executive agent for destroying CWM. Nonstockpile CWM is being handled by the Non-Stockpile Chemical Materiel Program (NSCMP); stockpile CWM is the responsibility of the Chemical Stockpile Disposal Program.[2] Because nonstockpile CWM is stored or buried in many locations, the Army is developing transportable disposal systems that can be moved from site to site as needed. The Army has plans to test prototypes of three transportable systems—the rapid response system (RRS), the munitions management device (MMD), and the explosive destruction system (EDS)—for accessing and destroying a range of nonstockpile chemical agents and militarized industrial chemicals. The RRS is designed to treat recovered chemical agent identification sets (CAIS), which contain small amounts of chemical agents and a variety of highly toxic industrial chemicals. The MMD is designed to treat nonexplosively configured chemical munitions (i.e., munitions containing chemical agents but no fuzes, propellants, or burster charges). The EDS is designed to treat munitions containing chemical agents with energetics equivalent to three pounds of TNT or less. These munitions are considered too unstable to be transported or stored. A prototype EDS system has recently been tested in England by nonstockpile program personnel. Although originally proposed for evaluation in this report, no test data were available to the committee on the composition of wastes from the EDS. Therefore, alternative technologies for the destruction of EDS wastes will be discussed in a supplemental report in fall 2001. Treatment of solid wastes, such as metal munition bodies, packing materials, and carbon air filters, were excluded from this report.

Because of differences in the solvents and chemical agents in CAIS materials and recovered chemical munitions, the RRS and MMD use different neutralization chemistries and produce different liquid waste streams—collectively referred to in this study as "neutralent wastes" or "neutralents." A summary of nonstockpile CWM that will be treated by the RRS and MMD, as well as the major constituents of their neutralent waste streams, is given in Table ES-1. According to the Army, the maximum permissible concentration for blister agents in a neutralent stream is 50 parts per million (ppm) (although in practice the actual concentration is more likely to be about 1 ppm). The maximum for nerve agents is 20 to 30 parts per billion (ppb). RRS neutralents may contain arsenic, a toxic heavy metal that must be captured and immobilized.

Because neutralent wastes from the RRS and MMD are expected to be classified as hazardous wastes under the

---

[1] The Convention on the Prohibition of the Development, Production, Stockpiling, and Use of Chemical Weapons and Their Destruction, known as the Chemical Weapons Convention, was signed by the United States on January 13, 1993, and ratified by the U.S. Congress on April 25, 1997. The CWC specifies deadlines for the destruction of CWM covered by the treaty. Countries may apply for an extension of the deadline of up to five years.

[2] The *stockpile CWM* (the subject of the Army's Chemical Stockpile Disposal Program) consists of both bulk containers of nerve and blister agents and munitions, including rockets, mines, bombs, cartridges, projectiles, and spray tanks, loaded with nerve or blister agents. CWM located at stockpile sites (i.e., stockpile CWM) will be disposed of during destruction campaigns at those sites.

TABLE ES-1 Transportable Treatment Systems and Neutralent Waste Streams Considered in This Study

System	Type of Non-Stockpile Chemical Materiel Treated	Key Constituents of Neutralent Waste Streams	Percentage by Weight
Rapid Response System (RRS)	Chemical Agent Identification Sets (sulfur mustard, nitrogen mustard, lewisite)	chloroform	50–84
		t-butyl alcohol	0–27
		water	0–2.4
		hydantoin derivatives	1–6
		various organics	0–9
		arsenic	not available
Munitions Management Device (MMD)	Nonstockpile chemical munitions without explosive components (sulfur mustard, phosgene, VX, GB)	water	7–90
		monoethanolamine[a]	34–90
		sodium hydroxide	4.2–9
		various organics	0–9
		various trace metals	not available

[a]Not used in the treatment of phosgene
Source: Adapted from U.S. Army, 1999a.

Resource Conservation and Recovery Act (RCRA),[3] the Army's current plan is to send them to a permitted hazardous waste incinerator for final disposal. However, the incineration of chemical agents elsewhere has aroused considerable opposition among some public interest groups, and this opposition may be extended to the incineration of RRS and MMD neutralents (even though the concentration of agent in the neutralents will range from ppb to a few ppm). In anticipation of increasing public opposition, the Army is investigating alternative (nonincineration) technologies for disposing of neutralents and has asked the National Research Council for advice. This report is a result of that request.

STATEMENT OF TASK FOR THIS STUDY

The following Statement of Task was given to the National Research Council by the Army:[4]

> Evaluate the near-term (1999–2005) application of advanced (nonincineration) technologies, such as from the Army's Assembled Chemical Weapons Assessment Program and the Alternative Technologies and Approaches Project, in a semi-fixed, skid-mounted mode to process Rapid Response System, Munitions Management Device, and Explosive Destruction System liquid neutralization wastes.

Around the time the committee was conducting this study, the Army asked two other contractors to undertake similar, though not identical, studies. Mitretek was asked to evaluate the applicability of six technologies being investigated by the Assembled Chemical Weapons Assessment (ACWA) Program (part of the stockpile CWM). Stone & Webster was asked to publish a Commerce Business Daily announcement requesting proposals for alternative technologies for the destruction of neutralents and to evaluate the proposals received. The committee received briefings on these projects and took account of them in its deliberations.

COMMITTEE'S APPROACH

The committee began by establishing some boundaries for the study. As required by the Statement of Task, only liquid neutralent wastes from the RRS and MMD were considered. First, EDS neutralents were omitted because the liquid neutralent (at the time this report was developed) had not been well characterized. Second, the end point of the neutralent treatment technology was taken to be solids that could be disposed of in a permitted landfill and liquids that could be released to a federally owned or publicly owned treatment works. Third, the air discharges would contain only CO_2, water vapor, and nitrogen. Therefore, setting discharge parameters would not be necessary.

The committee's approach to identifying technologies with the greatest potential for the timely, cost-effective treatment of RRS and MMD neutralents consistent with the protection of human health and the environment had two aspects. Clearly, legacy equipment developed by the ACWA Program and mature commercial destruction technologies that have the potential to destroy RRS and MMD neutralent, do not involve incineration, and require little or no development investment should be considered first. The committee recognizes that the Army is not starting its selection process with a blank slate. Several alternative technologies have

---

[3]Under RCRA, a substance is determined to be a hazardous waste either because it is listed as such in the law (a listed hazardous waste) or because its characteristics meet the conditions specified in the law for a hazardous waste (e.g., corrosivity).

[4]The original contractual language was updated and modified in discussions with the Army, resulting in the Statement of Task that follows.

already been initially evaluated as part of the ACWA Program and in commercial projects for treating hazardous wastes. If RRS and MMD neutralents could be effectively destroyed by "piggybacking" on ACWA or mature, commercial destruction technologies, (e.g., wet-air/O_2 oxidation [WAO], chemical oxidation, or PLASMOX®[5] as used today for waste disposal purposes) this might provide a relatively inexpensive and expedient course of action.

In the event that none of the existing ACWA legacy equipment or commonly used commercial technologies in their present form and state of development prove to be acceptable, the committee assembled a list of alternative treatment technologies that might, with development and investment, meet the needs of the NSCMP. These technologies became the focus of the committee's investigations and analyses. Eight candidate technologies were evaluated based on the collective judgment of the committee that the technology is likely to be safe, effective, and permitted, as well as consistent with pollution prevention principles. These eight technologies were ranked in order of preference:

1. chemical oxidation
2. wet-air/O_2 oxidation (WAO)
3. electrochemical oxidation with silver Ag(II) and cesium Ce(IV)
4. supercritical-water oxidation (SCWO)
5. solvated-electron technology (SET)
6. gas-phase chemical reduction (GPCR)
7. plasma-arc technology
8. biodegradation

Because the neutralents of the RRS and the MMD are very different chemically, the committee assessed the appropriateness of each technology for each type of neutralent separately. The ability of a given technology to destroy both types of neutralent effectively was considered a plus, but no technology was rejected if it would be effective for only one neutralent stream.

The committee's criteria were based on best practices in the chemical industry, including a criterion based on pollution prevention (e.g., minimizing the volume of material that must be added at the front end of the process and minimizing the production of high-temperature vapor streams at the back end of the process). The committee divided the best practices into two categories: top priority criteria and important criteria. The top priority criteria are:

- inherent safety
- technical effectiveness
- pollution prevention
- permit status

The important criteria are:

- robustness
- cost
- practical operability
- continuity
- space efficiency
- materials efficiency

Unlike the Mitretek and Stone and Webster studies mentioned above, the committee made no attempt to assign quantitative weights to the criteria. Instead, the technologies were evaluated qualitatively. Because the Army had no information on actual tests of the destruction of real or simulated nonstockpile neutralents, the committee relied on the expert judgment of committee members to evaluate each process and to suggest the most promising technologies for development.

Although the committee's primary objective was to evaluate alternative treatment technologies for neutralent waste streams, the committee also took into account public and regulatory acceptability, which are likely to affect the selection of alternative technologies. Some public interest groups opposed to incineration who have been actively involved in the policy debate were invited to attend committee meetings. As often as feasible, committee members met with public interest groups and quasigovernmental citizen committees to solicit their views, which were also considered in the committee's evaluations. In addition, the committee incorporated information from discussions with the Environmental Protection Agency (EPA) on potential regulatory approaches for expediting the implementation of alternative technologies for treating neutralent waste.

FINDINGS AND RECOMMENDATIONS

Technical Issues

Finding. The committee did not find any experimental studies on the destruction of neutralent wastes generated by the RRS or MMD. Therefore, the analyses of candidate technologies are based on their demonstrated performance with chemically similar materials, as well as on fundamental principles of chemistry and chemical engineering.

Finding. Based on the amount of neutralent expected from planned operations at Deseret Chemical Depot and Dugway Proving Ground, the volume of neutralents generated by the RRS and MMD is expected to be relatively small—on the

---

[5] Although the PLASMOX® process has not been permitted in the United States, it is in use in Switzerland for commercial applications and is being investigated by the Army. The committee notes that, because the PLASMOX® process uses oxygen, it is difficult to consider it as an alternative process to incineration.

order of 5,000 gallons per year in normal operation. As a point of reference, a standard tanker truck contains 5,000 to 10,000 gallons, and a railcar may contain as much as 30,000 gallons. Because the facility for disposing of neutralent will not have to handle large volumes or have a high throughput, it could be a laboratory or pilot-plant scale facility. Thus equipment for technologies currently under investigation for stockpile CWM might be used cost effectively for treating nonstockpile neutralents. At this small scale, all of the technologies reviewed by the committee could be adapted to "semi-fixed, skid-mounted" configurations (see Statement of Task).

Finding. The committee identified some low-temperature, low-pressure, less complex technologies that might be used to treat neutralent waste. The benefits of these technologies over incineration include low worker risk, public acceptance, low risk to the surrounding community, and simplicity of operation.

Finding. The Army's evaluation of alternative technologies must meet the time constraints of the CWC, which requires that all nonstockpile CWM in storage at the time the convention was ratified be destroyed by 2007. Thus far, no alternative incineration technologies have been tested on real, or even simulated, nonstockpile neutralent generated by either the RRS or the MMD. Therefore, bench testing and scale-up demonstrations of candidate technologies with neutralents will be necessary. Because testing the effectiveness of alternatives and determining regulatory limits will take time, the Army may have to fall back on its current incineration strategy for the destruction of neutralent, which includes the use of commercial incinerators, or even the use of the Army's stockpile incinerators.

Finding. Some of the candidate alternatives to incineration for destroying MMD and RRS neutralents involve hardware that has already been developed, and using them would simply require substituting neutralent for existing feeds. For example, one or more of the demonstration units tested for the chemical disposal programs (e.g., ACWA Program) might be used. Because the volume of nonstockpile neutralents will be small, even if the technology is not rated highly according to the committee's criteria but is inherently safe, the savings in time and development costs might justify consideration of this alternative. Demonstration units could be used at their present sites or moved, either as needed or to a mutually agreeable location based on a plan developed with the affected communities and regulatory authorities.

Recommendation. The Non-Stockpile Chemical Materiel Program should pursue a two-track strategy similar to the one adopted by the committee during its selection of a technology: (1) an evaluation of the potential of Assembled Chemical Weapons Assessment demonstration technologies

and mature commercial technologies; and (2) technologies that would require further development and investment.

Recommendation. As part of the track-one strategy, the Army should take advantage of available equipment that would require little or no investment (i.e., either alternative technologies from the Assembled Chemical Weapons Assessment [ACWA] Program or existing commercial technologies, such as chemical oxidation, wet-air/O_2 oxidation, or PLASMOX®). The following technologies from the ACWA demonstrations should be considered: electrochemical oxidation Ag(II), gas-phase chemical reduction, solvated-electron technology, and supercritical water oxidation. If any of these can accomplish the task safely, it might provide a relatively rapid and inexpensive course of action.

Recommendation. If Assembled Chemical Weapons Assessment (ACWA) or the commercial technologies require substantial modifications to processes or permits, the Army should focus first on the most easily adaptable commercial technologies (i.e., chemical oxidation and wet-air/O_2 oxidation). Only if these technologies prove to be unsuitable should the Army consider investing resources in the further development of ACWA technologies (listed below in order of preference):

- electrochemical oxidation with Ag(II) and Ce(IV)[6]
- supercritical water oxidation
- solvated-electron technology
- gas-phase chemical reduction
- plasma-arc technology

Recommendation. The Army should not invest in further development of biodegradation, which was judged least likely to be effective.

Regulatory Issues and Public Involvement

Recent experience by federal agencies has shown that the involvement of diverse public groups (including state and federal regulators) is crucial to timely decision making. Stakeholder involvement is particularly important for decisions involving analytical, engineering, or other scientific uncertainties about the protection of human health and the environment. The Army's implementation of an alternative technology or technologies to incineration could be delayed unless regulatory requirements have been developed and the public has been involved in the decision-making and selection process.

The NSCMP could improve its existing public involvement program by (1) exploring ways to ensure representation

---

[6]Although not an ACWA technology, this variant of electrochemical oxidation, Ce(IV), should be evaluated.

of diverse public groups in assessments of disposal technologies and associated regulatory issues; and (2) by working closely with potential host communities to identify and address their concerns.

A comprehensive regulatory compliance plan that involves all stakeholders could be essential to the timely implementation of an alternative technology. An environmental criteria working group, with representatives of the Army, the Environmental Protection Agency, state regulators, officials of the U.S. Department Health and Human Services, public interest groups, and citizens at large, could be formed to undertake advanced planning with the goals of (1) ensuring that substantive regulatory requirements can be met and (2) determining if additional testing or evaluations will be necessary to satisfy public or regulatory concerns.

Finding. Citizens groups that met with the committee strongly urged that the Army consider the long-term storage (i.e., longer than one year) of neutralents rather than incineration. Storage, they argued, would ensure that the Army would have sufficient time to develop, test, and obtain regulatory approval of alternatives to incineration. The committee believes that the Army's mission could be affected by the manner in which it responds to these public concerns.

Finding. The Army provided several reasons for not storing neutralent. First, storage might make it impossible to meet the treaty deadlines for the destruction of the nonstockpile chemical weapons. Second, the Army might be required to meet rigorous, long-term environmental requirements. Third, long-term storage would be inconsistent with regulatory requirements limiting storage time for hazardous wastes. Finally, the cost of storage might be disproportionately high.

Recommendation. To solicit public understanding, and perhaps acceptance, in its decision on whether or not to store neutralent, the Army should issue a detailed white paper explaining the legal, scientific, regulatory, and institutional issues involved. The paper should explicitly describe how risk to the public and workers would be affected by the long-term storage of neutralent prior to its disposal.

Finding. The committee's discussions with citizen groups indicated a need for, and the value of, public involvement in the Army's decisions concerning the selection, deployment, and employment of technologies for disposing of nonstockpile chemical materials.

Recommendation. The committee recommends that the Army expand its public involvement program regarding disposal of nonstockpile chemical materiel. Enough time should be scheduled and enough resources allocated to ensure that the decision-making process is open and that members of the public are involved in determining trade-offs related to the selection, siting, deployment, and employment of disposal technologies.

1

Overview

Since World War I, the United States has considered it necessary to have the capability to engage in chemical warfare. Today, however, chemical warfare materiel (CWM) accumulated over the years is considered obsolete and dangerous, and the United States and other signatories of the Chemical Weapons Convention (CWC) are committed to destroying all recovered CWM by 2007.[1]

U.S. law and international treaties have divided CWM into two categories: stockpile and nonstockpile materiel. Stockpile materiel includes all chemical agents available for use on the battlefield, including chemical agents assembled into weapons and chemical agents stored in bulk (one-ton) containers. Stockpile materiel is stored at eight locations in the United States and on Johnston Island in the Pacific Ocean.

Nonstockpile materiel includes all *other* chemical weapon-related items, such as buried CWM, recovered CWM, binary chemical weapons, former production facilities, and miscellaneous materiel. Much of the CWM was buried on military sites but is being rediscovered as the land is returned to the civilian sector. Some CWM is also buried at former test and firing ranges. According to the CWC, nonstockpile CWM items in storage at the time of treaty ratification (April 1997) must be destroyed within two, five, or ten years, depending on the type of chemical weapon and the type of agent. Nonstockpile CWM recovered after treaty ratification must be declared and destroyed "as soon as possible" (U.S. Army, 1999a).

The Army's Program Manager for Chemical Demilitarization (PMCD) has overall responsibility for disposing of all CWM under PMCD's two programs: the Chemical Stockpile Disposal Program and the Non-Stockpile Chemical Materiel Program (NSCMP). Although this study is concerned with the destruction of nonstockpile materiel, a brief review of the Chemical Stockpile Disposal Program is given below for two reasons. First, this program has been in progress for a longer time than the NSCMP. Second, many of the technologies and social and political factors that have influenced the Chemical Stockpile Disposal Program are expected to influence the NSCMP.

CHEMICAL STOCKPILE DISPOSAL PROGRAM

Baseline Program

In November 1985, Congress passed Public Law 99-145, which requires the destruction of stockpile agents and munitions. Therefore, the U.S. program to destroy stockpile chemical materiel was well under way at the time the CWC was first signed (January 1993). The Army selected incineration as the baseline method for destroying chemical agent in the stockpile materiel; two incinerators, one on Johnston Atoll in the Pacific Ocean and one at the Deseret Chemical Depot near Tooele, Utah, are currently in operation. Together these incinerators are expected to destroy about one-half of the U.S. stockpile, the remainder of which is dispersed among seven storage sites in the continental United States.

Because federal law (P.L. 103-337) prohibits the interstate shipment of chemical weapons, the Army had planned to construct similar incineration systems at the seven other sites. In fact, baseline facilities have been permitted and are under construction at three sites: Anniston, Alabama; Pine Bluff, Arkansas; and Umatilla, Oregon.

---

[1] The Convention on the Prohibition of the Development, Production, Stockpiling and Use of Chemical Weapons and Their Destruction, known as the Chemical Weapons Convention, was signed by the United States on January 13, 1993, and ratified by the U.S. Congress on April 25, 1997. The CWC specifies deadlines for the destruction of CWM covered by the treaty. Countries may apply for an extension of up to five years.

Alternative Technologies for Destroying the Stockpile

Because incineration as a disposal technology has met with strong public and political opposition, the Army began a search for alternative, nonincineration technologies for destroying stockpile chemical agents in two key programs: one for chemical agents stored in bulk, one-ton containers (sulfur mustard [HD] at Aberdeen Proving Ground, Maryland, and VX [a nerve agent] at Newport, Indiana). In addition, as directed by Congress, the Army is investigating alternate disposal technologies for chemical agents in assembled chemical weapons at two other sites (Pueblo, Colorado, and Lexington Blue Grass, Kentucky) (NRC 1999b).

Alternative Technologies and Approaches Program

In April 1994, the Department of the Army issued the *Alternative Demilitarization Technology Report to Congress* (U.S. Army, 1994a). A subsequent report (U.S. Army, 1994b) outlined an aggressive research and development program to evaluate two alternatives to incineration: neutralization alone and neutralization followed by biodegradation.

In 1994, the Army's Product Manager for Alternative Technologies and Approaches (ATAP), under the Office of Chemical Demilitarization, undertook a focused research program on the proposed neutralization-based processes for agent destruction. As a result, two processes have been developed. Hydrolysis of chemical agent in pure water followed by biodegradation has been developed to destroy the HD stored in bulk containers at Aberdeen, Maryland. Neutralization by aqueous sodium hydroxide, followed by supercritical water oxidation (SCWO), is being developed to destroy the VX stored in bulk containers at Newport, Indiana. The Army is in the process of designing, constructing, and testing these neutralization-based systems (NRC 1994, 1998a, 2000a).

Alternative Technologies Program for Assembled Chemical Weapons Assessment

In 1996, Congress also appropriated money and mandated that the Army demonstrate at least two nonincineration technologies for the destruction of assembled chemical weapons at Pueblo, Colorado, and Lexington Blue Grass, Kentucky. The Army established the Assembled Chemical Weapons Assessment (ACWA) Program (NRC, 1999b) to carry out this mandate. Seven technologies passed the initial screening. One was eliminated shortly thereafter because of technical problems. Three were selected for demonstration (Demo I): plasma-arc technology, hydrolysis followed by treatment with SCWO, and hydrolysis followed by biodegradation (NRC, 2000a). Prototype equipment for unit operations was constructed and tested, and engineering design is under way for integrated systems.

Congress subsequently mandated that ACWA also test the remaining three undemonstrated technologies: electrochemical oxidation, gas-phase chemical reduction (GPCR), and solvated-electron technology (SET). Demonstrations for these three technologies were started in early summer 2000 and ended in September 2000 (Demo II). Results of these demonstrations were received too late for inclusion in this report. Following Demo II, the Army will determine whether alternative technologies will be used at the Pueblo, Colorado, and Lexington Blue Grass, Kentucky, stockpile sites.

NON-STOCKPILE CHEMICAL MATERIEL DISPOSAL PROGRAM

Prior to 1991, efforts to dispose of CWM were limited to stockpile materiel. A part of the 1991 Defense Appropriations Act (House Appropriations Report 101-822) directed the Secretary of Defense to establish an office with the responsibility of destroying nonstockpile materiel. The program manager for NSCMP was assigned this task under the newly established U.S. Army Chemical Materiel Destruction Agency (NRC, 1999a).

Nonstockpile Sites

In the 1993 Defense Appropriations Act (P.L. 102-484, Section 176), Congress directed the Army to (1) report the locations, types, and quantities of nonstockpile chemical materiel; (2) specify the methods to be used for its destruction; (3) provide cost and time estimates; and (4) assess transportation options. In a *Survey and Analysis Report*, the Army provided an overview of its task (U.S. Army, 1996). According to this report, nonstockpile CWM is located at more than 200 sites in the United States and U.S. territories. CWM at most sites includes small quantities of chemical agent but does not appear to pose immediate hazards to the public or the environment. However, chemical weapons agreements and continuing discoveries of contaminated sites have increased the impetus for locating and disposing of all nonstockpile CWM.

The purpose of the NSCMP is to provide centralized management and direction for the characterization and destruction of nonstockpile CWM, develop disposal facilities, provide schedule and cost estimates, and ensure compliance with federal, state, and local regulations.

Transportable Treatment Systems

Because a large number of locations have only small quantities of CWM, the Army decided to develop transportable disposal systems that can be moved from site to site as needed. To treat the entire range of materiel (e.g., munitions containing a variety of chemical agents,[2] some configured

---
[2]See Table 1-1 in *Disposal of Chemical Agent Identification Sets* (NRC 1999a).

TABLE 1-1 Transportable Treatment Systems for Nonstockpile Chemical Materiel

System	Type of Materiel	Status
Rapid Response System (RRS)	Chemical Agent Identification Sets (CAIS).	Full-scale prototype designed and assembled; testing was completed summer 2000, however, the results were not available for inclusion in this report.
Munitions Management Device (MMD)[a]	Nonstockpile chemical munitions with no explosive components; small containers of chemical agent; chemical samples.	Full-scale prototype designed and assembled; testing was completed summer 2000, however, the results were not available for inclusion in this report.
Explosive Destruction System (EDS)	Chemical munitions with explosive components.	Phase I prototype in testing; Phase II model in design.

Source: U.S. Army, 1999b.

[a]The development of a successor system, the MMD-2, designed to handle explosively configured materiel, has been suspended for cost reasons. Explosively configured materiel will probably be treated in a fixed facility to be built in Pine Bluff, Arkansas, where most of the recovered munitions are currently stored (Brankowitz, 2000).

with explosives), the Army decided it would need three separate transportable systems (Table 1-1). Two of the three systems—the rapid response system (RRS) and the munitions management device (MMD)—are ready for testing. The RRS is designed to treat chemical agent identification sets (CAIS). The MMD is designed to treat nonexplosively configured munitions and containers filled with sulfur mustard, HD, phosgene, sarin (GB), and VX. These two systems are the focus of this study. The third system, the explosive destruction system (EDS), has been tested in England, but the neutralent waste has not yet been characterized.

Both the RRS and MMD systems can be mounted on a series of trailers, and both use chemical processes to treat agents. The design of an MMD-2 system, intended to treat explosively configured munitions, has recently been postponed (Brankowitz, 2000). At the time of this writing, the RRS and MMD systems had been permitted for testing in Utah.

Handling Processes

Figure 1-1 is a flow chart illustrating the Army's planned disposition of nonstockpile CWM. As the figure shows, when CWM is discovered, CAIS items are separated from munitions and other containers and treated in the RRS, producing a neutralent waste stream (inside the dotted box). Munitions are then evaluated to determine the type of chemical agent fill and whether they contain energetics. Chemical agent fill is analyzed by using portable isotopic neutron spectroscopy (PINS). If the munitions do not contain energetics, they are treated in the MMD, which also produces a neutralent waste stream (inside the dotted box).

Munitions containing energetics with a total explosive force of about one pound of dynamite (e.g., 75-mm projectiles, 4.2-inch mortar rounds, 8-inch live rounds) are expected to be treated in the EDS Phase 1. Munitions containing larger quantities of explosives will be treated in the EDS Phase 2, which is currently being designed and is expected to be accepted by mid-2002. The waste streams shown inside the dotted box at the bottom of Figure 1-1 are the subject of this study. Waste streams outside the box, as well as alternative treatment processes that might replace the RRS, MMD, and EDS, will be considered in another study next year.

Neutralent Waste Streams

In both the RRS and MMD, munitions or containers are opened, and liquid reagents are mixed with the chemical agents. According to Army test data, the agent concentration in the reaction vessel is thereby reduced to less than 50 parts per million (ppm) for mustard and lewisite and less than 50 parts per billion (ppb) for VX and GB (U.S. Army, 1999b). However, because of the treatment reagents, the liquid waste streams from both the RRS and MMD (called "neutralents" in this study) will contain chlorinated organic chemicals (only RRS), excess reactants, and reaction products and are likely to be considered hazardous waste under Subtitle C of the Resource Conservation and Recovery Act (RCRA), which provides national standards (regulations) for the "cradle-to-grave" management of hazardous waste. Thus, RRS and MMD neutralents cannot be released directly to the environment unless they are treated further. In addition, neutralents may contain Schedule 2 precursors,[3] which also

---

[3]Under the CWC, Schedule 2 chemicals have limited commercial utility and can be readily converted to chemical weapons. Production of these chemicals above specified limits is subject to reporting requirements and verification through on-site inspections.

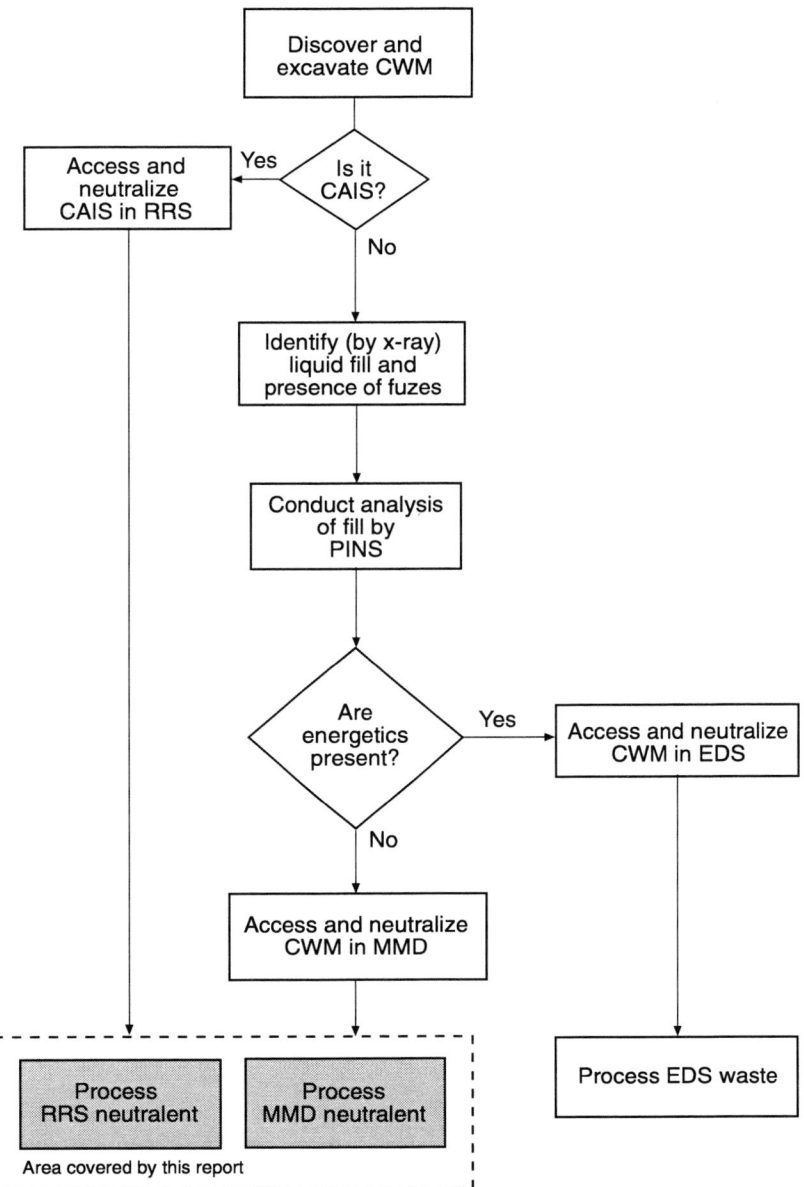

FIGURE 1-1 Flow chart for the disposal of nonstockpile CWM in transportable systems. Waste streams in the dotted box are the focus of this report.

require further treatment. Currently, although the Army plans to incinerate neutralent derived from its mobile systems, alternative destruction technologies are also being investigated.

Studies of Alternative Technologies

The Army enlisted two organizations to study alternatives to incineration for the treatment of neutralents. Mitretek was asked to determine whether technologies originally proposed for the treatment of stockpile CWM under the ACWA Program might also be used to treat neutralents generated from the RRS and MMD. Using weighted evaluation criteria, Mitretek ranked the six ACWA technologies in order of their suitability for treating the neutralents.[4]

---

[4]Mitretek ranked the six technologies in the following order of preference: (1) SCWO, (2) GPCR, (3) plasma arc technology, (4) silver II, (5) SET, and (6) hydrolysis/biotreatment (Mitretek, 1999).

The Army also requested that Stone and Webster publish a Commerce Business Daily announcement calling for proposals for technologies to treat RRS and MMD neutralents and then evaluate their applicability. Stone & Webster's analysis[5] is limited to the proposals received in response to the announcement.

APPLICABILITY OF ACWA TECHNOLOGIES

All six of the technologies submitted to the ACWA Program have been tested, although the results of the most recent tests (Demo II), which were completed in September 2000, have not yet been published. Much of the equipment used in the demonstration tests will continue to be used by the Army, and three of the test units would be large enough to treat the total quantities of MMD and RRS neutralents. All of the technologies selected for testing were considered acceptable alternatives to incineration by the citizen stakeholders involved in the selection process. Because the equipment used in the ACWA demonstrations may be available to treat neutralents in the near future, using the demonstration equipment may prove to be an expedient and cost-effective solution for the destruction of neutralents even though the technology (assuming it meets safety criteria) is not the one best suited to the job.

Regulatory Requirements

The use of the RRS and MMD mobile systems presents significant regulatory challenges. Many federal and state regulations will affect the treatment and management of neutralents. The entire process, including neutralization, storage, transportation, and ultimate disposal of the neutralent, will be regulated under Subtitle C of RCRA. The Clean Water Act may apply, if the neutralent is sent to a publicly owned treatment works (POTW).

RCRA regulatory requirements are codified under federal and state rules and regulations. In some cases, state environmental regulations are even more stringent than federal regulations, making a single national approach practically impossible. Additional regulatory requirements may apply if some of the chemicals in the neutralents are defined as "lethal chemical agents" under the CWC and U.S. statutes (e.g., 50 U.S.C. § 1512 (j)(2)).

ROLE OF THE NATIONAL RESEARCH COUNCIL

Involvement with the Non-Stockpile Chemical Materiel Disposal Program

This is the second of three reports the National Research Council was asked to produce for the Army. In the first report, delivered in 1999, the committee reviewed disposal options for CAIS (NRC, 1999a). In this second report, the committee evaluates nonincineration technologies for the treatment and/or destruction of neutralents. However, due to a recent change in the strategic direction of the NSCMP, the National Research Council was asked to produce two additional reports (in place of the third report) over the ensuing 18 months. The first, scheduled to be delivered in late 2001, will make recommendations for disposing of EDS waste streams; the second, to be delivered in early 2002, will evaluate alternative (nonincineration) strategies for the comprehensive treatment of nonstockpile CWM, including agents, energetics, and munitions/containers.

Statement of Task for This Study

The committee was given the following Statement of Task for this study:[6]

> Evaluate the near-term (1999–2005) application of advanced (non-incineration) technologies, such as from Army's Assembled Chemical Weapons Assessment (ACWA) program and the Alternative Technologies and Approaches Project (ATAP), in a semi-fixed, skid-mounted mode to process Rapid Response System (RRS), Munitions Management Device (MMD), and Explosive Destruction System (EDS) liquid neutralization wastes.

SCOPE OF THIS STUDY

This report focuses on nonincineration, alternative technologies for the treatment of liquid neutralents from the RRS and MMD; the method by which the agent is accessed and separated from the munition or container is not considered. Solid waste streams from the RRS and MMD (e.g., carbon filters, metal parts, dunnage) are not considered. The treatment of liquid waste streams from the EDS, which have not been well characterized but will most likely contain both unexploded energetics and by-products from the explosives used to access the munitions, will be considered in a separate report.

The nonincineration technologies considered in this study are based on the following sources:

---

[5]Stone & Webster's technology evaluation panel recommended the following six technologies: (1) catalytic hydrothermal conversion technology; (2) catalytic transfer hydrogenation technology; (3) gas-phase chemical reduction; (4) MGC PLASMOX® process; (5) solvated-electron/persulfate oxidation technology; and (6) supercritical water oxidation (Stone & Webster, 2000).

[6]The original contractual language was updated and modified in discussions with the Army, resulting in the Statement of Task that follows.

- ACWA Program demonstrations
- ATAP demonstrations
- proposals received by Stone and Webster (or the Army) in response to the Commerce Business Daily announcement
- previous reviews by the North Atlantic Treaty Organization (NATO)
- knowledge and personal experience of committee members

Technologies are discussed generically without reference to particular vendors, although the committee obtained information from some specific technology providers to complete its evaluation. In these cases, every effort was made to treat the technology as generically as possible.

Each technology was evaluated for its capability to meet the following goals:

- Solids wastes can be sent to a RCRA Subtitle C or D landfill.[7]
- Any residual agents, Schedule 2 compounds, or other materials in the neutralents are at low levels that do not preclude discharge directly to a POTW or federally owned treatment works (FOTW).
- Air discharges contain only CO_2, water vapor, oxygen, and nitrogen. (Note that the degree of containment, capture, and processing required to achieve those conditions could vary greatly.)

Technologies that could meet these goals would require no further treatment except for the normal biodegradation that takes place at a POTW. Whether or not a POTW will, in fact, accept nonhazardous wastewater derived from the treatment of chemical agents has not been determined.

During the course of the study, the committee was informed by the Army that treatment, storage, and disposal facilities (TSDFs) had been surveyed as potential disposal sites. Only TSDFs whose primary technology was incineration responded. However, in wet-air/O_2 oxidation (WAO), one of the most promising technologies for the destruction of neutralent, the committee suggests that the Army also investigate TSDFs that use this technology.

Because no information was available on actual tests of the destruction of real or simulated nonstockpile neutralent, or even any paper studies, the committee had to rely on the expert judgment of committee members to predict the most likely outcomes for each technology and identify the most promising technologies for development. Because the RRS and MMD neutralents are chemically very different from each other, the committee assessed the appropriateness of technologies for each waste stream separately. If a given technology could effectively destroy both types of neutralents, this was considered an advantage, but no technology was rejected if it would be effective for only one neutralent stream.

The committee also attempted to include the views of the interested public in its deliberations. The committee met with federal regulators from the Environmental Protection Agency's (EPA) Office of Waste Management and with members of public interest groups who have been active in the policy debate. The latter were invited to present and discuss their views with the committee and to participate in site visits to observe two of the technologies being evaluated. In addition, members of the committee observed meetings of a stakeholder group convened by NSCMP (called the CORE Group) that included both regulators and members of the public.

Committee Approach

The committee adopted a dual approach to evaluating and selecting the technologies with the greatest potential for treating RRS and MMD neutralents in a timely, cost-effective manner consistent with the protection of human health and the environment. The first approach was opportunistic. The committee reasoned that, if the neutralents could be effectively destroyed by "piggybacking" on existing stockpile or other hazardous waste destruction campaigns, this might provide a relatively inexpensive, time-efficient, and convenient solution, even if the technology was not rated highly for destroying the neutralents according to the committee's criteria.

The second approach was based on best practices in the chemical industry, which are documented in public records. The committee ranked the selected technologies qualitatively according to these criteria. Unlike the Mitretek and Stone and Webster studies mentioned above, the committee made no attempt to weight the criteria quantitatively.

STRUCTURE OF THIS REPORT

The neutralents generated from the RRS and MMD mobile treatment systems, which are the "input" to the disposal technologies considered in this study, are characterized in some detail in Chapter 2. Chapter 3 describes the derivation of the criteria the committee used to evaluate the technologies. Chapter 4 presents brief descriptions of the selected alternative technologies, along with the committee's evaluations and rankings. Chapter 5 discusses issues related to public involvement in the technology selection process. Chapter 6 presents the committee's findings and recommendations.

---

[7] A RCRA Subtitle C landfill accepts hazardous wastes; a RCRA Subtitle D landfill accepts municipal solid wastes. The committee believes that residual solids from the processes discussed in this report can be stabilized and pass regulatory requirements for disposal. However, treatability studies will be necessary to demonstrate this.

2

Waste Streams from Transportable Treatment Systems

The NSCMP is preparing to test two prototype transportable systems, the RRS and the MMD, to destroy a range of nonstockpile chemical agents and militarized industrial chemicals. The RRS is designed to treat recovered CAIS, which were used to train soldiers in the detection and identification of chemical agents and decontamination procedures. The MMD is designed to treat nonexplosively configured chemical munitions (i.e., munitions containing chemical agents but no fuzes, propellants, or burster charges).

Because CAIS and recovered munitions contain different materials, the RRS and MMD must use different reagents to destroy the chemical agents. Although both systems appear to be effective for destroying the target agents (to ppm levels or below), they also produce liquid waste streams containing complex mixtures of reaction by-products, excess reagents, and organic solvents. These liquid waste streams are referred to in this study as "neutralents."[1]

Because neutralent waste streams from the RRS and MMD are expected to be classified as hazardous wastes under RCRA,[2] the Army had planned to ship them to a permitted hazardous waste incinerator for final disposal. However, because the incineration of chemical agents has aroused considerable opposition among public interest groups, and because this opposition may be extended to the incineration of neutralents, the Army is also investigating alternative (nonincineration) technologies for disposing of neutralents.

This chapter describes the composition, quantity, and toxicology of the neutralent waste streams expected to be generated by the RRS and MMD. In keeping with the Statement of Task for this portion of the study, the committee accepted the treatment processes and neutralent compositions as given. The committee did not consider upstream changes in the treatment chemistry or process conditions that might produce neutralents with different characteristics.

CHEMICAL AGENT IDENTIFICATION SETS

Approximately 110,000 CAIS were produced in various configurations from about 1928 to 1969. These sets contain (1) neat chemical agents and/or (2) agents dissolved in chloroform in glass vials or glass bottles and/or (3) agents adsorbed on charcoal in glass bottles. The chemical agents include blister agents, sulfur mustards (HD and H), nitrogen mustard (HN-1 and HN-3), and lewisite.[3]

RAPID RESPONSE SYSTEM

Treatment Processes

The treatment chemistry of the RRS is based on the oxidation of chemical agents with 1,3-dichloro-5,5-dimethylhydantoin (DCDMH) dissolved in a mixture of chloroform/t-butyl alcohol/water. This reagent was selected because it does not react with chloroform, which is present in some CAIS items, it maintains a reasonable reaction volume, and it does not generate significant amounts of heat or gaseous products during the reaction.

Depending on the CAIS item and the agent it contains, one of four reaction processes is selected. The processes (blue, red, charcoal, and charcoal-L) have different proportions

---

[1] The RRS and MMD also produce solid waste streams that include metal munition bodies, packaging materials, and carbon air filters, but these are not included in this study.

[2] Under RCRA, a substance is determined to be a hazardous waste either because it is listed as such in the law (a listed hazardous waste) or because its characteristics meet the conditions specified in the law for a hazardous waste (e.g., corrosivity).

[3] CAIS also contain a variety of highly toxic industrial chemicals, such as phosgene, but in the RRS process these are identified, repackaged, and sent to a commercial incinerator for disposal. Thus, they do not contribute to the RRS waste stream and are not considered further here.

TABLE 2-1 CAIS Chemical Agents and Treatment Processes in the Rapid Response System

Chemical Agent	Treatment Reagent	Percentage by Weight	Process Designation
Nitrogen mustard (HN-1), (HN-3), sulfur mustard (HD), and lewisite (L) in chloroform solution	chloroform t-butyl alcohol water DCDMH	58.5 30.3 2.4 8.8	Red
Neat sulfur mustard (H)	chloroform t-butyl alcohol water DCDMH	58.5 30.3 2.4 8.82	Blue
Nitrogen mustard (HN-1, HN-3) and sulfur mustard (HD) adsorbed onto charcoal	chloroform DCDMH	89.0 11.0	Charcoal
Lewisite adsorbed onto charcoal	chloroform t-butyl alcohol water DCDMH	58.5 30.3 2.4 8.8	Charcoal-L

Source: Gieseking, 2000.

of DCDMH and different solvents (see Table 2-1). The blue process is used to treat neat sulfur mustards (H or HD). The red process is used to treat sulfur mustard, nitrogen mustard (HN-1, HN-3), and lewisite that are dissolved in chloroform. The charcoal process is used to treat sulfur mustards and nitrogen mustards adsorbed on charcoal, and the charcoal-L process is used to treat lewisite adsorbed on charcoal.

Neutralent Waste Streams and Volumes

The chemical reactions of the RRS processes are complex, and a large number of products are present in the neutralent waste streams. However, all four processes effectively destroy the chemical agents and produce waste streams that could be shipped to hazardous waste incinerators. The compositions of the neutralent waste streams are shown in Table 2-2. The dominant constituents consist of the reaction solvents and excess DCDMH. A large number of reaction by-products (e.g., sulfones, sulfoxides, etc.) are present in low concentrations.

Arsenic found in lewisite is converted in the red and charcoal-L processes into chlorovinylarsonic acid (CVA) in quantities of up to 3 percent by weight in the neutralent waste stream (Table 2-2). The fate of CVA depends on the post-treatment processes. In a SCWO reactor and in the GPCR caustic scrubber brine, the CVA is expected to be converted to sodium arsenate salts (e.g., Na_3AsO_4 and $Na_4As_2O_7$), which can then be treated with ferric chloride to produce ferric arsenic salts for disposal in a hazardous waste landfill. This treatment scheme, which was developed in Canada, is the basis of the treatment of bulk lewisite at the Chemical Agent Munitions Disposal System facility in Utah.

If 40 to 45 CAIS ampoules or 12 to 15 CAIS bottles are treated per day, the estimated volume of neutralent generated would be less than 15 gallons per day (U.S. Army, 1999a). The 1,189 CAIS items located at Deseret Chemical Depot are expected to generate a total of about 468 gallons of liquid neutralent (Gieseking, 1999).[4]

MUNITIONS MANAGEMENT DEVICE

Treatment Processes

The treatment chemistry of the MMD is based on (1) the hydrolysis of HD and GB with monoethanolamine (MEA) and water or (2) the hydrolysis of VX with MEA-aqueous sodium hydroxide solution. MEA was chosen as the reagent based on previous experience with it in Russian chemical demilitarization programs. The advantages of MEA include good solvent properties for agents, miscibility with water, noncorrosivity to stainless steel under operating conditions, and low flammability. MEA cannot be used in the RRS because it reacts violently with chloroform, the solvent present in many CAIS items.

---

[4]This estimate is an upper boundary, based on the assumption that all CAIS items contain agent. If CAIS items that contain industrial chemicals are repackaged and sent to a hazardous waste incinerator for disposal, the volume of neutralent waste could be reduced by about half.

TABLE 2-2 Composition of Neutralent Waste Streams from the Rapid Response System[a]

Waste Component	Blue Process (percentage by weight)	Red Process (percentage by weight)	Charcoal or Charcoal-L Process (percentage by weight)
Chloroform	54.5–55.5	60–61	50–84
t-butyl alcohol	26–27	17–20	0–24
Water	2.2–2.4	1.7–1.9	0–1
Dichlorodimethyl hydantoin unreacted DCDMH		0–4.6	0–7
Chlordimethyl hydantoin (CDMH)	2.1–5.9	1.9–5.6	2–6
5,5 dimethyl hydantoin (DMH)	1–3	0–4.6	0–3
Chlorinated sulfoxides (diethyl and ethylvinyl)	5.4–7.6	0.6–2.1	0–0.4
Chlorobutanes and chlorobutenes	2.4–3.4	1.2–4.6	0–4
Chlorinated sulfones (diethyl and ethylvinyl)	0–0.1	0–0.06	0–0.3
1,1,2 trichloroethane	0–0.015	0–0.23	0–0.025
Tetrachloroethane[b]	0–0.025	0–0.2	0–0.022
Bis-(2-chloroethyl) amine		0–1	0–0.5
Chlorovinylarsonic acid		0–2.6	0–3
Acetaldehyde and chloroacetaldehyde		0–0.5	
Polychlorinated diethyl sulfides and polychlorinated ethylvinyl sulfide			0–2
Dichloroethane[c]			0–0.03
Pentachloroethane			0–0.03
Hexachloroethane[c]			0–0.01
Chloral hydrate			0–0.7
Glass/plastic	2–3	7.5–10	5–8
Charcoal			5–5.2

Note: Waste composition includes other organics, such as carbon tetrachloride; 1,1 dichloroethylene; tetrachloroethylene; trichloroethylene; and vinyl chloride. Waste composition also includes toxic characteristic metals, such as arsenic, barium, cadmium, chromium, lead, mercury, nickel (not a TCLP constituent, but listed in Appendix VIII—Hazardous Constituents in 40 CFR 261), selenium, and silver. All metals may not be present in all wastes. Lewisite contains arsenic. Data on concentrations are not yet available for either organics or metals.

[a]RCRA characterization of the neutralent waste stream will be completed using analytical data obtained from bench-scale demonstrations conducted at the Edgewood Chemical and Biological Center, Aberdeen Proving Ground, Maryland.
[b]May be either isomer, 1,1,1,2-tetrachloroethane, or 1,1,2,2-tetrachloroethane.
[c]RCRA toxic characteristic leaching procedure (TCLP) constituents.

Source: Adapted from U.S. Army, 1999a.

Depending on the type of agent in the munition, one of three reagents is selected for the MMD process (see Table 2-3): a mixture of MEA and water; MEA and aqueous sodium hydroxide; or just aqueous sodium hydroxide. Phosgene is reacted with aqueous sodium hydroxide to form simple inorganic salts. A minimum ten-fold volume excess of reagent solution is used to ensure destruction of the agent and to control viscosity.

Neutralent Waste Streams and Volumes

The compositions of neutralent waste streams from the MMD are complex, as shown in Tables 2-4 to 2-7. Because many of the munition bodies introduced into the MMD have already degraded during their long burial, metals and debris may be mixed with the neutralent during the cutting and rinsing processes. If a large amount of sodium hydroxide is added during processing, the pH of the neutralent may exceed 14 (hazardous waste).

In current testing of the MMD, one CWM munition or container can be processed per day. In future tests or in normal operation, the rate may be two items per day (U.S. Army, 1999a). The initial testing of the MMD at Dugway Proving Ground in Utah is expected to generate approximately 6,412 gallons of neutralent, an average of 57 gallons of liquid waste (including liquid wastes from rinsing the system after processing) for every gallon of agent or industrial chemical processed (Gieseking, 1999).

TOXICITY OF NEUTRALENTS

A number of reports produced or sponsored by the Army describe dermal and inhalation toxicity studies of the oxidant/solvent systems (O/SS) used in the RRS and MMD,

TABLE 2-3 Reagents Used to Neutralize Chemical Agents in the MMD

Chemical Agent or Industrial Chemical	Treatment Reagent	Percentage by Weight
Sulfur mustard (HD)	MEA	90
	water	10
Sarin (GB)	MEA	45
	water	55
Nerve gas (VX)	MEA	86
	water	7
	sodium hydroxide	7
Phosgene	water	90
	sodium hydroxide	10

Source: Gieseking, 2000.

as well as a waste stream from these systems (e.g., U.S. Army Research Office, 1994; DOT, 1997; and U.S. Army, 1999a). The effects of exposure to individual components of the O/SS are shown in Table 2-8. In general, the toxicity (i.e., exposure response data) of both RRS and MMD neutralents is comparable to the toxicity of components of the O/SS (U.S. Army, 1999a).

Neutralent from the Rapid Response System

The inhalation toxicity of neutralents from the RRS red process treatment of HD, HN, and lewisite was tested in rats by 14-day exposures. The neutralent contained 53 percent chloroform, 30 percent t-butyl alcohol, trace amounts of DCDMH, and less than 1 ppm HN or HD, or 37 ppm lewisite. The toxicity of the waste stream was compared with that of an aerosol containing 58.3 percent chloroform, 39.1 percent tert-butanol, and 2.6 percent water (the vehicle control). Concentrations of 24,000 ppm of the vehicle control or neutralent killed all of the test animals. Lower doses caused excessive salivation, ocular and nasal discharge, lack of coordination, listlessness, difficult breathing, and corneal opacity. The inhalation effects of the neutralent on test animals were consistent with those of the t-butanol and chloroform components of the O/SS (Morgan et al., 1997).

The dermal toxicity of RRS neutralents from all four processes and O/SS was tested by exposing the skin of rabbits to the solutions under an occluded patch for 24 hours. All solutions caused redness and swelling, but with the exception of the charcoal process, the effects of the neutralent were less severe than those caused by the O/SS. The dermal effects of the charcoal process neutralent were comparable to those of the O/SS because of the moderately toxic HD degradation products produced when DCDMH reacts with HD in the absence of water (DOT, 1997).

The vesicant (blister formation) properties of the neutralents from RRS processes were tested by dermal application to hairless guinea pigs. The only neutralent that caused vesication was from the blue process (treatment of neat HD). Because the concentration of HD in neutralent (less than 50 ppm) is too low to cause vesication, the blistering was attributed to the presence of HD oxidation products (Olajos et al., 1997).[5]

Neutralent from the Munitions Management Device

Unlike the reaction between DCDMH and HD in the RRS, reaction between MEA and HD in the MMD produces relatively few toxic breakdown products. This is reflected by the lower toxicity of the HD neutralents from the MMD process.

The dermal toxicity of O/SS and simulated neutralents from the treatment of HD, GB, and VX were compared by exposing the skin of rabbits to the solutions under an occluded patch for four hours. The effects were recorded 24 hours after exposure. Severe redness and swelling were observed in all cases, but for the most part, skin injuries from the neutralents and O/SS alone were comparable. No systemic toxicity resulted from 24-hour dermal exposures to either the O/SS or neutralent solutions (Olajos et al., 1996).

The vesicant properties of the O/SS and the HD/MEA/water neutralent were studied by dermal application to hairless guinea pigs. Blisters did not result from dermal exposure to either the O/SS or the neutralent solution (Battelle, 1997).

FEDERAL AND STATE REQUIREMENTS

Because neutralents contain compounds that are classified as hazardous, they will be regulated under RCRA Subtitle C. They will also be regulated under the CWC based on the chemical agent(s) they contain and under the U.S. Department of Transportation (DOT) regulations because both hazardous materials and chemical agents have special requirements for transport in the United States.

RCRA Subtitle C is the "cradle-to-grave" approach of managing hazardous waste, including generation, storage, shipment, treatment, and disposal. Under RCRA, neutralents produced by the RRS and the MMD may be classified either as listed or characteristic hazardous wastes. If the neutralent waste stream contains phosgene, it will either be classified as a listed hazardous waste, or, if it is corrosive (pH >10), as

---

[5]HD reacts with DCDMH to form sulfoxides, which are relatively nontoxic. However, they can react with excess DCDMH to form sulfones, which have vesicant properties comparable to those of HD.

TABLE 2-4 Composition of Sarin (GB) Neutralent Wastes from Bench-Scale Tests of the MMD

Waste Component	Concentration
Major Constituents	
Water	49.4–49.0 wt %
Monoethanolamine (MEA)	33.9–40.3 wt %
2-hydroxyethylammonium O-isopropyl methylphosphonate salt	0.7–8.5 wt %
Monoethanolamine hydrofluoride salt	0.4–4.6 wt %
O-isopropyl O-(2-aminoethyl)methylphosphonate	0.3–3.0 wt %
Minor Constituents	
Diisopropyl methylphosphonate (DIMP)	0.03–0.36 wt %
Tributylamine (TBA)	0.2–0.017 wt %
1,3-diisopropylurea (DIPU)	45–530 ppm
1,3-diisopropylthiourea (DIPTU)	17–200 ppm
2-hydroxyethylammonium methylphosphonate salt	400–800 ppm
Other methylphosphonates	< 100 ppm
Sarin (GB)	ND (< 25 ppb)
RCRA TCLP Constituents	
Organics	
Benzene[a]	6.5–6.8 mg/l
Hexachlorobutadiene[b]	1.0–1.6 mg/l[c]
2,4-dinitrotoluene[c]	0.2–1.6 mg/l[c]
Hexachlorobenzene[c]	0.2–1.6 mg/l[c]
Methyl ethyl ketone (MEK)[d,e]	0.29–0.54 mg/l[c]
Metals	
Arsenic[d]	0.66–0.76 ppm
Barium[d]	ND–0.75 ppm
Chromium[d]	410–1080 ppm
Lead[d]	550–1300 ppm
Nickel[e]	410–500 ppm

Note: Treatment reagent percentage by weight: water (55 percent), MEA (45 percent).
[a]RCRA toxicity-characteristic component concentration greater than TCLP regulatory level.
[b]RCRA toxicity-characteristic components. Quantitation limits were above TCLP regulatory limits.
[c]Source: Dugway Proving Ground, 1998.
[d]RCRA toxicity-characteristic component concentration less than TCLP regulatory limit
[e]Not a TCLP constitutent. Included because it is listed in Appendix VIII—Hazardous Constituents in 40 CFR 261.

Source: Adapted from U.S. Army, 1999a.

a characteristic hazardous waste (40 CFR 261; NRC, 1999a).[6] The neutralent waste stream could be regulated under the federal or state requirements (or both) of RCRA. In some cases, state requirements are more stringent than federal requirements.

---

[6]CFR citations refer to the U.S. Code of Federal Regulations with the volume number preceding CFR and the section number following. Copies of volumes of the U.S. Code of Federal Regulations are available through the Government Printing Office outlets and commercial document and regulatory services.

The storage of GB and VX neutralents from the MMD is subject to the constraints of the CWC. Some of the breakdown products (e.g., amiton [S-[2-(diethylamino)ethyl-phosphorotioic acid O,O-diethyl ester]), are listed as Schedule 2 precursors to the manufacture of chemical agents (i.e., chemicals that could be used to remanufacture chemical agent). This means that, theoretically, the precursor chemicals in the neutralent could be reprocessed from the neutralent and used to remanufacture chemical agents. To prevent the manufacture of chemical weapons, the CWC requires that Schedule 2 precursors derived from existing

TABLE 2-5 Composition of Mustard (HD) Neutralent Wastes from Bench-Scale Tests of the MMD

Waste Component	Concentration
Major Constituents	
Monoethanolamine (MEA)	67–89 wt %
Water	8.9–9.9 wt %
Monoethanolamine hydrocloride	0.9–13.8 wt %
N-(2-hydroxyethyl)thiomorphooline (HETM)	0.6–9.1 wt %
Bis-[(2-hydroxyethylamino)ethyl] sulfide (HEAES and other organic sulfides)	0.05–1 wt %
Minor Constituents	
1,4-dithiane	0.008–0.16 wt %
Chlorinated thiophenes	< 1 [a]
Mustard (HD)	ND (< 50 ppb)
RCRA	
Organics	
Tetrachloroethylene[b]	2.2–2.6 mg/l
Trichloroethylene[b]	1.4–1.6 mg/l
Vinyl chloride[b]	5.8–6.9 mg/l
Hexachlorobutadiene	2.0–3.3 mg/l [a]
2,4-dinitrotoluene	2.0–3.3 mg/l [a]
Hexachlorobenzene	2.0–3.3 mg/l [a]
1,1-dichloroethylene [c]	0.13–0.15 mg/l
Chloroform[c]	0.14–0.2 mg/l
Methyl ethyl ketone (MEK)[c,d]	0.33–0.37 mg/l
Metals	
Arsenic[c]	0.14–0.23 ppm
Chromium[c,d]	0.531–0.62 ppm
Nickel[d]	0.13–0.15 ppm
Selenium[b]	3.0–3.6 ppm

Note: Treatment reagent percentage by weight: water (10 percent), MEA (90 percent).
[a]Source: Dugway Proving Ground, 1998.
[b]RCRA toxicity-characteristic component concentration greater than TCLP regulatory level.
[c]RCRA toxicity-characteristic component concentration less than TCLP regulatory limit.
[d]Not a TCLP constituent. Included because it is listed in Appendix VIII—Hazardous Constituents in 40 CFR 261.

Source: Adapted from U.S. Army, 1999a.

agents be destroyed in the same time frame as the chemical agents.

The reader should note that Tables 2-4 to 2-7 describing the composition on neutralent wastes derived from bench tests of the RRS and MMD neutralents do not list any Schedule 2 precursor compounds. Yet the Army suggests that a major argument against storage is that neutralents may contain Schedule 2 compounds and therefore must be destroyed per the CWC schedule. The committee does not find this to be inconsistent. Whenever chemical agents are treated, Schedule 2 breakdown products could be produced. The bench test data in the tables indicates that this did not occur in these tests. However, until more is known about compounds produced by the reactions occurring in the RRS and MMD and a significant body of data has been established for large-scale operation, it is best to take a conservative approach and assume that some Schedule 2 breakdown products will be present in sufficient quantity to preclude storage.

Significant concerns have been raised about the transport of CWM. Indeed, concern about the movement of chemical agents has been a driving force behind the development of

TABLE 2-6 Composition of VX Neutralent Wastes from Bench-Scale Tests of the MMD

Waste Component	Concentration
Major Constituents	
Monoethanolamine (MEA)	77.6–83.0 wt %
Water	6.9–7.0 wt %
Sodium hydroxide	4.2–6.3 wt %
Sodium 2-diisopropylaminoethanethiolate (NaThiol)	1.4–0.5 wt %
Sodium O-ethylmethylphosphonate (NaEMPA)	0.6–2.0 wt %
Sodium O-(2-aminoethyl) methylphosphonate (NaAEMPA)	0.5–1.8
Minor Constituents	
Disodium methylphosphonate (Na$_2$MPA)	0.15–0.5 wt %
Bis-2(-diisopropylaminoethyl)sulfide (Sulfide)	0.22–0.71 wt %
Bis-2(-diisopropylaminoethyl)disulfide (Disulfide)	0.13–0.41 wt %
2-diisopropylaminoethyl ethyl sulfide	0.03–0.09 wt %
1,3-dicyclohexylurea	0.1–0.35 wt %
Ethanol	0.2–0.7 wt %
Unquantified identified products[a]	0.4–1.0 wt %
VX	ND (< 1 ppm)
RCRA TCLP Constituents	
Organics:	
Benzene[b]	1.0–7.5 mg/l[c]
Carbon tetrachloride[b]	< 1.0 mg/l[c]
1,2-dichloroethane[b]	< 1.0 mg/l[c]
1,1-dichloroethane[b]	< 1.0 mg/l[c]
Tetrachloroethane[b]	< 1.0 mg/l[c]
Trichloroethane[b]	< 1.0 mg/l[c]
Vinyl chloride[b]	< 1.0 mg/l[c]
Metals:	
Chromium[d]	0.38–0.44 ppm
Lead	1.2–1.4 ppm
Selenium	< 1.0–4.1 ppm

Note: Treatment reagent percentage by weight: MEA (86 percent), water (7 percent), sodium hydroxide (7 percent).

[a]Compounds identified: cyclohexylamine (CHA); 2-disopropylamino ethanol (DIPAE); 2-diisopropylamino ethanethiol (VX thiol); 2-(diisopropylamino)ethyl sulfide (DIPAES); chloromethyl-2-(diisopropylamino) ethyl sulfide (DIPAMS); N-2[(chloromethylthio) methylthio]ethyl-N-isopropyl-2-propanamine; bis(2-diisopropylaminoethyl)sulfide (VX sulfide); bis(2-diisopropylaminoethyl)disulfide (VX disulfide); N-2-O[(2-diisopropylamino)ethylthiomethylthio ethyl-N-isopropyl-2-propanamine (VX Me disulfide); ethylene glycol (EG); N-2-hydroxyethyl methylphosphoramidate (VX-N-MEA).
[b]RCRA toxicity characteristic component concentration greater than TCLP regulatory level.
[c]Source: Dugway Proving Ground, 1998.
[d]RCRA toxicity-characteristic component concentration less than TCLP regulatory limit.

Source: Adapted from U.S. Army, 1999a.

transportable treatment systems. As far back as 1969 (P.L. 91-121), Congress placed severe, almost insurmountable, restrictions on the transport of CWM, including a requirement for advance notification and coordination of shipments with the U.S. Department of Health and Human Services and Congress (except in cases of emergency). In 1995, Congress placed restrictions on moving nonstockpile CWM out of any state except to the closest permitted CWM storage facility, and then only under very strict conditions. Public concerns about transporting CWM have effectively foreclosed even this option except in extraordinary situations.

TABLE 2-7 Composition of Phosgene Neutralent Wastes from Bench-Scale Tests of the MMD

Waste Component	Percentage by Weight
Water	90
Sodium hydroxide (NaOH)	8–9
Sodium carbonate (Na$_2$CO$_3$)	1–2
Sodium chloride (NaCl)	1–2

Note: Treatment reagent percentage by weight: water (90 percent), sodium hydroxide (10 percent).

Source: Adapted from U.S. Army, 1999a.

TABLE 2-8 Toxicity of Components of the O/SSs Used in the RRS and MMD

Oxidant/Solvent System Component	Inhalation Toxicity	Dermal Toxicity	Eye Contact
t-butanol	High concentrations cause incoordination and narcosis.	Slight skin irritation.	Severe irritation.
Chloroform	Central nervous system depression; toxic to liver and kidneys; classified as probable human carcinogen.	Repeated or prolonged exposure causes irritation and defatting.	High vapor concentrations cause conjunctivitis and spasmodic winking; contact with liquid causes a burning sensation and reversible injury to the corneal epithelium.
Dichloro-dimethyl-hydantoin (DCDMH)	Severe irritation of respiratory tract; high concentrations can cause difficulty breathing and pulmonary edema.	Severe irritation.	Severe irritation.
Monoethanolamine (MEA)	Irritation of the respiratory tract.	Severe irritation.	Severe irritation.

Source: U.S. Army, 1999a.

The neutralents generated by the RRS will primarily include hazardous waste and hazardous materials, which make them subject to RCRA and DOT requirements. If these are the only constituents, the neutralent could be transported as a routine hazardous waste or hazardous material under existing laws and regulations as long as it was properly packaged, marked, manifested, and shipped as required by those regulations.

However, the neutralent from the RRS and MMD may contain trace amounts of residual chemical agents (e.g., GB [< 25 ppb], sulfur mustard [< 50 ppm], VX [< 1 ppm]). At these levels, the toxicity studies cited above indicate that the concentration of residual agent would be too low to affect the overall toxicity of the waste streams. It is not known how much, if any, chemical agent would be present at concentrations lower than the detection limit. Thus, based on the information provided by the Army, transporting neutralent wastes from the RRS and the MMD should not be subject to any restrictions beyond the applicable federal RCRA, DOT, and state regulatory requirements. However, the public perception of "residual chemical agents" from the MMD waste streams may arouse concerns.

Although these residues will be in extremely small amounts, the public could consider the overall neutralent waste mixture as "tainted" with chemical agent and, therefore, of special concern. The Army should address this potential problem proactively. This could be done in several ways. In a previous report, for example, the committee recommended that a comparative risk assessment be performed of the disposal of CWM in CAIS in an incinerator and the disposal of typical hazardous waste (NRC, 1999a). The Army could assess the comparative risk of transporting and disposing of neutralent and transporting and disposing of typical hazardous waste. Providing the public with this type of information would increase the transparency and credibility of the process.

3

Criteria for Evaluating Technologies

A great many nonincineration technologies could theoretically be used to treat the neutralents from the MMD and the RRS. The committee selected the most promising technologies (see Table 3-1) from the following sources:

- recent NATO reviews, which include many general descriptions
- previous studies by the National Research Council of the destruction of chemical agents
- current programs by the Army and Army contractors for the destruction of agents or neutralent
- commercial experience with the destruction of other waste streams

TEMPERATURE CLASSIFICATIONS

The technologies were classified into four categories according to operating temperature: low temperature, moderate temperature, high temperature, and very high temperature.

The technologies classified as **low temperature** operate at temperatures of less then 100°C (boiling point of water). These technologies do not require pressurized containment. Technologies classified as **moderate temperature** operate at temperatures of 100°C to 370°C. Technologies classified as **high temperature** operate at temperatures of 370°C to 1,000°C. Technologies classified as **very high temperature** operate at temperatures of more than 1,000°C.

PRESSURE CLASSIFICATIONS

The technologies were classified into four categories according to operating pressure: low pressure, moderate pressure, high pressure, and very high pressure. The technologies classified as **low pressure** operate at pressures of less than 15 pounds per square inch absolute (psia).[1] These technologies do not require pressurized containment for an aqueous system when processing waste. A wide range of reasonably standard support equipment (e.g., pumps, flanges, valves, etc.) rated for up to 615 psia are available. Technologies operating at pressures from 15 to 615 psia were classified as **moderate pressure**. Above 615 psia but still in Division I (less than 3,015 psia), considerably more care is necessary in the design of the process vessels to prevent leakage. Technologies operating at pressures of 615 to 3,015 psia were classified as **high pressure**. Above 3,015 psia (Divisions II and III), designing process vessels required specific individual designs and calculations, as well as special requirements for support and containment structures. Technologies operating at pressures of more than 3,015 psia were classified as **very high pressure**.

SELECTION CRITERIA

The committee did not have the time or resources to evaluate all of the technologies. Therefore, only the most promising technologies were selected for more detailed evaluation. These technologies were selected according to the following criteria:

- If a great deal of information was available, and the technology was under serious consideration and/or evaluation for other demilitarization or waste treatment purposes (e.g., technologies being tested under the ACWA Program), it was selected for evaluation.

---

[1] Fifteen (15) psia is the established transition point between low-pressure tank and pressure vessel design standards covered under the API Std. 620 and ASME Boiler & Pressure Vessel Code, Section VIII (Div. I for under 3015 psia, Div. II & III for over 3015 psia).

TABLE 3-1 Technologies Selected for Evaluation

Technologies	Oxidation or Reduction	Temperature	Pressure
Chemical oxidation	oxidation	low	low
Biodegradation	oxidation	low	low
Electrochemical oxidation (Ag(II) and Ce(IV))	oxidation	low	low
Solvated-electron technology (SET)	reduction	low	moderate
Wet-air /O_2 oxidation (WAO)	oxidation	moderate	moderate to very high
Supercritical water oxidation (SCWO)	oxidation	high	very high
Gas-phase chemical reduction (GPCR)	reduction	high	low to moderate
Plasma-arc technology	oxidation	very high	low to moderate

- If, in the committee's collective judgment, a technology was likely to be safe, effective, and permitted, and also likely to rate satisfactorily on the pollution prevention criteria (see below), it was selected for evaluation, and efforts were made to gather more information.

The eight technologies selected cover a broad range and are not limited to the technologies advocated by technology providers.

TOP PRIORITY CRITERIA

Relatively Safe Processes (Low Risk)

Technologies were reviewed to determine if a common process failure (e.g., explosion, corrosion, mechanical failure, operator error, incorrect feeds, service failure, etc.) under normal operating conditions could lead to serious worker, community, or environmental damage. The following factors were considered:

- minimal storage and transportation of hazardous materials
- minimal toxicity and flammability of all materials
- temperatures and pressures below the threshold values that challenge reliable containment

Technical Effectiveness

Technologies were evaluated for their consistency in achieving a standard (in this case, destruction) of neutralent. The following factors were considered:

- efficiency of detoxification of the neutralent (i.e., solid wastes could be disposed of in a landfill and liquid wastes released to a POTW)
- integration into a system for the destruction of nonstockpile materiel

Permit Status

Technologies were evaluated for serious regulatory obstacles that would prevent environmental and/or operational permitting. The following factors were considered:

- potential major delays in obtaining permits under federal (and international), state, or local regulations
- potential for meeting schedules of international treaties

Pollution Prevention

The committee evaluated the technologies on the principle of "green chemistry" (Mulholland and Dyer, 1999). In other words, pollution prevention and waste minimization practices are implemented at the beginning of the process (pollution prevention) as opposed to after the fact (pollution abatement). The following factors were considered:

- minimal addition of processing materials[2] that would require treatment, disposal, regeneration, recycling, or other handling
- minimal number of processing steps, which all have an incremental environmental burden in potential leaks and energy, maintenance, shutdown and start-up, and clean-out requirements

---

[2]Processing materials include not only the obvious purchased solvents, acids, bases, etc., and service materials, such as catalysts, filters, and adsorbents, but also common items, such as water, nitrogen for instruments and vapor-space inerting, and nitrogen in air used as a source of oxygen. These materials might be used for the process itself or for support tasks, such as cleaning.

- minimal toxicity of emissions, wastes, or other material that require treatment, disposal, regeneration, recycling, or other handling[3]
- operating temperatures and pressures as close to ambient as possible
- minimal corrosion, plugging, sensitive process-control parameters, and other operating difficulties
- minimal high-temperature vapor streams that require high-quality treatment

IMPORTANT CRITERIA

Once the selected technologies had been evaluated according to top priority criteria, they were evaluated by the important criteria.

Robustness

A robust technology can function successfully in stable continuous operation. The term "continuous" means the technology can treat neutralent from beginning to end and does not require another technology as an intermediate step before final disposal. Continuous also means that feedstock can be continuously supplied or supplied in the batch mode. Operation of a robust technology has the following characteristics:

- tolerance of normal variations (differences in concentrations of hazardous materials or chemical agents)
- start-up and shutdown of a facility without major complications or delays
- operation at small scale or large scale, as required
- capability of treating a wide range of potential feeds (neutralents from the RRS and MMD)

Cost

Although the committee did not conduct a cost analysis for each technology, cost was estimated based on past experience and knowledge. The following cost factors were considered:

- total costs, including capital and operating costs
- costs per unit of feed

Practical Operability

The following factors related to practicality were considered:

- minimal training for operators (average skill levels for the chemical industry)
- use of standard instrumentation for monitoring and process controls

Continuity

Two factors were considered in this category:

- likelihood of finding a vendor
- likelihood that supplies of raw materials will be available

Space Efficiency

The main factor in space efficiency was the weight, area, and volume of operating equipment per volume of material processed.

Materials Efficiency

The following factors were considered:

- recycling of materials as part of the internal operation of the facility
- shipment of wastes off site for beneficial reuse
- use of recycled materials from external sources

Areas of Special Concern

Because of the lack of empirical information on neutralent treatment, the committee's approach to establishing evaluation criteria for the eight selected technologies was necessarily qualitative. Some particular areas of concern are included in those criteria that were not identified separately. These areas of concern are discussed in the write ups of specific technologies in Chapter 4 and are identified below:

- acetic acid (a compound resulting from oxidation processes that is difficult to oxidize further and will probably be present in neutralents; although easily biodegradable, its presence is a good indicator of the need for discharge to a POTW)
- arsenic (including oxides and metallo-organic compounds)
- nitrogen oxides (including NO_x, N_2O)
- sulfur compounds (SO_x, H_2S)
- dioxins and furans
- cleanup, decontamination, and relocation of facility

---

[3] For example, arsenic, which is present in lewisite neutralent, is a semivolatile metal in a high-temperature process. The arsenic is released as a vapor and condenses in the gases as a very fine, hard-to-capture particulate. The 1999 EPA incinerator regulations added stringent emission limits for semivolatile metals, and incinerator operators are, therefore, very cautious about accepting wastes containing organo-arsenic compounds.

4

Descriptions and Evaluations of Technologies

The committee selected eight candidate technologies to evaluate: chemical oxidation, WAO, biodegradation, electrochemical oxidation [Ag(II) and Ce(IV)], SCWO, SET, plasma-arc technology, and GPCR. In this chapter, these technologies are described briefly (roughly in order of increasing operating temperature), evaluated, and ranked according to the criteria described in Chapter 3. None of these technologies has been tested on neutralents, and their effectiveness can only be estimated based on their use in similar applications. Experimental studies, including measurements of their destruction effectiveness on actual neutralents, will be necessary.

Each technology description is followed by tables representing the qualitative assessment of individual committee members assigned to investigate that technology based on their expertise. The committee took these qualitative assessments into account in its overall ranking of technologies.

CHEMICAL OXIDATION

On balance, chemical oxidation is a promising technology for mineralizing (i.e., converting organic compounds to inorganic salts, water, and carbon dioxide) RRS and MMD neutralents and for converting other components to less toxic materials. Experimental studies will be necessary to verify its effectiveness.

Description

Hydrogen peroxide, potassium permanganate, Oxone™,[1] peroxydisulfate, and ultraviolet (UV)-activated hydrogen peroxide or ozone oxidation are all viable oxidants for the treatment of nonstockpile neutralents.[2] Under appropriate operating conditions and with sufficient reagent, the organic compounds present in neutralents can be expected to be mineralized with any of these oxidants.[3]

For chemical oxidation not activated by UV light, conventional process equipment and procedures are used. The reactions are carried out at 80°C to 100°C at atmospheric pressure in aqueous solutions. When an organic phase is present, vigorous agitation is necessary to suspend and disperse the organic materials in the aqueous phase.

Processes employing UV activation[4] require special equipment. The solution containing the material to be oxidized must be pumped past a quartz tube containing a UV lamp to expose it to UV radiation (the solution must be transparent to UV radiation). If ozone is used, it must be generated in an ozone generator. The oxidation system is usually operated semicontinuously (i.e., a large batch of feed is prepared in a feed tank, pumped past the UV source, and returned to the feed tank). This operation is continued until the desired degree of oxidation is obtained. The contents of the feed tank are then discharged.

Evaluation

Chemical oxidation is a simple, well established industrial process that uses standard equipment under relatively mild conditions. The only gas evolved is carbon dioxide, and the aqueous reaction products can be evaporated to leave inorganic salts that can be stabilized and sent to a landfill. Because all reagents would be in aqueous solutions at

---

[1]Oxone, a registered trademark of DuPont Specialty Chemicals, is a triple salt ($2KHSO_5 \cdot KHSO_4 \cdot K_2SO_4$). The active component is $KHSO_5$, the potassium salt of monoperoxysulfuric acid (Cooper et al., 1999; DuPont Specialty Chemicals, 1992; Mikolajczyk, 1996).

[2]Palladium and other catalysts can also facilitate chemical oxidation.

[3]Oxidation with hypochlorite has been studied but does not appear to be as effective (Soilleaux, 1998).

[4]UV radiation is capable of decomposing ozone or hydrogen peroxide to form hydroxyl radicals that can oxidize most organic compounds (Holm, 1998). The hydroxyl radical has a high oxidation potential exceeded only by fluorine.

ambient pressure and below 100°C, dioxins and furans are not formed.

The cost of reagent is expected to be relatively high. Several pounds of reagent may be necessary to mineralize one pound of neutralent. However, because the total volume of neutralent will not be large, the cost of reagent is not expected to be an important consideration.

Because UV light cannot penetrate an opaque solution, the opacity of the feed will have to be considered for processes that include UV activation. Fouling and periodic cleaning of optical surfaces are design and operating considerations.

The biggest potential disadvantage of chemical oxidation is that it may not fully mineralize all of the compounds in the neutralents or that it may not mineralize them rapidly enough to be practical. This question can only be resolved through further research. The committee was not aware of any direct experience with the mineralization of neutralent by chemical oxidants. However, the oxidation or mineralization of closely related materials, including mustard, nerve agents, and their hydrolysates, has been documented. Laboratory-scale studies at the Army's Edgewood Research, Development and Engineering Center on the reaction of VX, GB, GD (soman), and

TABLE 4-1a Chemical Oxidation: Top Priority Criteria

Criterion	Rating
Technical Effectiveness	
Integral part of a coherent system	Good. All or parts of this technology could easily be integrated with existing nonstockpile treatment technologies.
Destruction efficiency	Potentially good, but must be verified by experiment. Complete mineralization if large quantities of chemical oxidant are used under correct conditions.
Inherent Safety	
Minimal storage and transportation of hazardous material	Good. Oxidizing agents must be transported and stored. Storage and transportation are routine.
Minimal toxicity and flammability (process materials)	Good. No highly toxic or flammable materials used.
Containable temperature and pressures	Fair. For hydrogen peroxide, conditions that minimize opportunities for decomposition must be used. The stability of hydrogen peroxide-containing reaction liquors depends on the concentration of the hydrogen peroxide, the temperature, and the materials present. Experimental studies will be necessary to determine the highest concentrations of hydrogen peroxide that can be safely employed.
Pollution Prevention	
Minimum toxicity of effluents	Good. Organic compounds can be effectively and rapidly destroyed. Under adequate conditions, nontoxic products, up to and including products of mineralization, can be produced. For hydrogen peroxide, this may require up to 3.5 pounds of peroxide per pound of organic compound. Chloroform can be mineralized.
Minimal use of processing materials	Poor. Large volumes of reagent may be necessary for oxidation, and large volumes of cement may be necessary for stabilization of residues.
Solids, liquids, and gaseous wastes	Fair. Carbon dioxide will be produced if mineralization is accomplished. A liquid waste stream will be produced, which can be converted to a solid via stabilization. No data have been generated on this topic, and treatability studies will be necessary.
Minimal number of processing steps	Good. May be a one-step or two-step process. Oxidation with or without stabilization is part of the process.
Temperatures, pressures, corrosion, plugging, and other operating difficulties minimized to prevent unprogrammed shutdowns	Good. Temperature and pressure are moderate. For hydrogen peroxide, conditions must be moderate.
Permit Status	
Allowed by regulations and capable of meeting schedules imposed by records of decision and treaties	Good. Permitting should be relatively easy.

mustard with hydrogen peroxide and peroxydisulfate had very favorable results (Hovanek et al., 1993; Yang, 1995, 1999). Using peroxydisulfate, VX was mineralized to carbon dioxide, nitrate, sulfate, and phosphate.

The use of hydrogen peroxide or Fenton's reagent was a key feature of the technology developed by ARCTECH and tested on hydrolysates for the ACWA Program. The procedures were shown to be effective at the bench scale for hydrolysates of VX, GB, and mustard. However, the ARCTECH technology was judged not to meet the demonstration selection criteria of the ACWA Program (see NRC, 1999b).

A chemical destruction process that uses base hydrolysis and oxidation with peroxydisulfate salts to mineralize chlorinated and other organic compounds has been developed at the Lawrence Livermore National Laboratory (Lawrence Livermore National Laboratory, 2000). Peroxidisulfate oxidation of MMD neutralents was proposed by Teledyne-Commodore in response to the Commerce Business Daily announcement promulgated by Stone and Webster. This process is likely to require a large excess of peroxydisulfate, leading to the formation of large quantities of sulfate in the waste stream (Yang, 1995).

UV/hydrogen peroxide oxidation (followed by biodegradation) was a feature of the Parsons/AlliedSignal technology for nerve agents demonstrated (unsuccessfully) for the ACWA Program (NRC, 1999b, 2000a). The combination of biological and UV/hydrogen peroxide treatment was able to achieve only 40 to 60 percent destruction of Schedule 2 compounds from GB hydrolysate, somewhat more for VX hydrolysate. The poor performance of the UV/hydrogen peroxide was attributed to the black color of the waste stream. The Gas Research Institute has conducted extensive studies on UV-enhanced ozone-based or hydrogen peroxide-based chemical oxidation of organic compounds associated with former gas plant sites (GRI, 2000).

Overall, chemical oxidation is a good candidate for treating neutralents because of its technical effectiveness, its good pollution-prevention qualities, its robustness, and its low cost. The technology can be easily integrated into existing nonstockpile treatment systems and is commercially used to treat other waste streams. Organic compounds can be destroyed at low temperature and pressure, and toxic emissions are minimal because no large gas streams, such as those encountered in combustion processes or in GPCR, are involved. Formation of chlorodibenzodioxins and chlorodibenzofurans is precluded because of the low temperatures. Based on commercial experience with chemical oxidation technology, capital and operating costs are expected to be moderate.

The ratings for chemical oxidation are summarized in Tables 4-1a and 4-1b.

TABLE 4-1b Chemical Oxidation: Important Criteria

Criterion	Rating
Robustness	
Stability and continuity of operation	Good, although chemical oxidation remains to be demonstrated. Stabilization of solids is not expected to cause problems.
Cost	
Minimal total costs	Good. Capital and operating costs are expected to be moderate.
Practical Operability	
Minimal training time	Good. Operations are conventional, and training should be similar to training for workers in chemical plants.
Continuity	
Vendor likely to remain in operation and raw materials likely to remain available	Unknown. No specific vendor.
Space Efficiency	
Minimal weight, area, and volume of operating equipment	Fair. Equipment, although conventional, may be large.
Materials Efficiency	
Use of internally recyclable materials	Unknown.
Beneficial reuse of wastes	Fair. Metals could be recycled, but not other residues.
Use of externally recyclable materials	Not applicable.

ELECTROCHEMICAL OXIDATION

The committee considered two forms of electrochemical oxidation, the silver Ag(II) process and the cerium Ce(IV) process (the "CerOx" process). The Ag(II) process has been advanced as a candidate for treating assembled chemical weapons but is probably not suitable for treatment of RRS neutralents because of their high chlorine content. Although Ag(II) and Ce(IV) are more potent oxidizers than the chemical oxidants discussed above, electrochemical processes are less desirable for treating neutralent wastes because they generate large quantities of hazardous effluents and because corrosive effluents could cause operating problems.

Ag(II) Process

Description

This process has been patented for oxidizing organic wastes using Ag(II), an unstable form of silver and one of the strongest oxidizing agents known. Any carbon in the waste stream is completely oxidized to carbon dioxide with traces of carbon monoxide. Other elements end up as salts (e.g., fluorines to fluorides, sulfur to sulfates). Chlorine precipitates out with the silver as silver chloride. The process is operated at 90°C and at atmospheric pressure.

A solution of silver nitrate in 8-molar nitric acid is electrolyzed to produce the Ag(II) cations at the anode of a commercially available electrochemical cell. A semipermeable membrane separates the anode and the cathode compartments of the cell to prevent mixing of the anolyte and catholyte solutions but allowing the passage of cations and water across the membrane.

The anolyte and catholyte solutions form two separate recirculating loops. The anolyte solution is circulated through the reaction vessel into which the organic wastes are introduced. Solids formed in the anolyte loop are removed by a hydrocyclone. In the cathode loop, the nitric acid is reduced to nitrous acid and water. This solution is passed through a nitrogen oxide reformer to regenerate nitric acid. Off-gases are passed through a scrubber. If no chlorine is present, the silver ions are recovered and recycled to the anolyte loop.

Evaluation

Ag(II) is expected to be an effective oxidizing agent for destruction of MMD neutralent. However, the large quantities of chloroform present in the RRS neutralent would result in the formation of large quantities of silver chloride, which would probably plug up the electrochemical cells.

The Ag(II) process also has several disadvantages. First, large quantities of concentrated nitric acid, which is extremely corrosive and a strong oxidizer, are required. Second, significant amounts of silver nitrate must be added (although, in principle, the silver is recovered in a recycling step), and silver is an expensive and regulated metal. Third, the large quantities of nitrogen oxides generated at the cathode must be reformed back to nitric acid, and waste gases must be scrubbed.

The Ag(II) process has been evaluated by the ACWA Program as an alternative technology for the disposal of assembled chemical weapons. The process was demonstrated with nerve agents, but not with mustard, at Porton Down in Great Britain. Two units have been constructed for the ACWA Demo II tests that were completed in September 2000—a small unit was tested with the chemical agent, and a large unit was tested with energetic materials. The results of these tests should be of interest to the NSCMP because either unit would be large enough to treat all of the neutralent generated from the MMD. The results of the tests were not available at the time this report was written.

The Ag(II) process as evaluated by the committee's criteria has both advantages and disadvantages. It has the technical capability to treat neutralent from the MMD but may be ineffective in treating neutralent from the RRS because of the large quantities of chloroform. The Ag(II) process has good space efficiency and stable continuous operation with reasonable controls, and the silver can be recycled. The major disadvantage for RRS neutralent is that large quantities of silver salts and chlorides are generated, which could lead to problems with corrosion and precipitation. Another disadvantage is that large quantities of silver (a toxic heavy metal) and nitric acid (a corrosive) are required for the operation of this technology, which could increase toxic emissions and effluents.

The ratings for electrochemical oxidation Ag (II) are summarized in Tables 4-2a and 4-2b.

TABLE 4-2a Electrochemical Oxidation Ag(II): Top Priority Criteria

Criterion	Rating
Technical Effectiveness	
Integral part of a coherent system	Fair. This process has not been interfaced with the neutralent process and has not been used to treat neutralent. The process is more attractive for treating MMD wastes than RRS wastes because of the high chlorine content in the latter.
Destruction efficiency	Good. Ag(II) has one of the highest known oxidation potentials.
Inherent Safety	
Minimal storage and transportation of hazardous material	Fair. Nitric acid is a hazardous material that is transported routinely. Large quantities of silver nitrate may be required.
Minimal toxicity and flammability (process materials)	Poor. Silver is a toxic heavy metal; nitric acid is highly corrosive.
Containable temperature and pressures	Excellent. Moderate temperatures (90°C), and pressures allow reliable containment.
Pollution Prevention	
Minimum toxicity of effluents	Good. Organic compounds can be effectively and rapidly destroyed. Under adequate conditions, nontoxic products, up to and including products of mineralization, can be produced. Chloroform can be mineralized.
Minimal use of processing materials	Fair. Large quantities of nitric acid and AgII are required.
Solids, liquids, and gaseous wastes	Poor. Large quantities of silver nitrate and nitric acid are required.
Minimal number of processing steps	Good. Small number of steps.
Temperatures, pressures, corrosion, plugging, and other operating difficulties minimized to prevent unprogrammed shutdowns	Fair. The system must be glass lined. Joints will have leakage and corrosion problems. Corrosion and precipitation could be serious problems, leading to plugging of the electrochemical cells.
Permit Status	
Allowed by regulations and capable of meeting schedules imposed by records of decision and treaties	Not known. Large quantities of silver salts, including silver chloride, will be produced, although the vendor intends to recycle all of the silver. Silver II has been operated at pilot scale at Dounreay and Porton Down, United Kingdom. There is no information as to whether or not it has been permitted as a full-scale facility.

TABLE 4-2b Electrochemical Oxidation Ag(II): Important Criteria

Criterion	Rating
Robustness	
Stability and continuity of operation	Fair.
Cost	
Minimal total costs	Fair. Requires an inventory of silver in the form of silver nitrate. The silver can be reconstituted by an outside firm.
Practical Operability	
Minimal training time	Fair. Training time is likely to be substantial.
Continuity	
Vendor likely to remain in operation and raw materials likely to remain available	Good.
Space Efficiency	
Minimal weight, area, and volume of operating equipment	Fair. Equipment, although conventional, may be large.
Materials Efficiency	
Use of internally recyclable materials	Good. Modular system was demonstrated at Aberdeen in spring 2000. Data are not yet available.
Beneficial reuse of wastes	Good. Silver will be recycled.
Use of externally recyclable materials	Not applicable.

CerOx Process

Description

The CerOx process is similar to the Ag(II) process except that it uses 0.8M Ce(IV) solution in 3-molar nitric acid at 100°C to oxidize and destroy organic compounds. Unlike Ag(II), Ce(IV) is stable. The Ce(IV) is produced and regenerated by the electrolysis of Ce(III) in a bipolar electrochemical cell, which the vendor calls a "T-cell."

The system has two circulating loops, one for the anolyte solution and one for the catholyte solution. In the anolyte loops, Ce(III) is oxidized to Ce(IV) in the T-cell and passed through the reaction chamber where the organic wastes are introduced gradually. Carbon is converted to carbon dioxide; chlorine compounds are converted to elemental chlorine, which is scrubbed and converted to hypochlorite; sulfur and other elements are converted to salts, such as sulfates. These salts remain in anolyte solution, which must be periodically replaced as the concentration of the salts increases.

The catholyte loop provides the second electrode for the electrolysis. The nitric acid in this loop is reduced to nitrous acid and then reformed back to nitric acid and nitric oxide. Water is produced in the process, but much of it is removed by evaporation because the operating temperature is very close to the boiling point (100°C).

The CerOx process uses very few reactants, principally nitrate (which is recycled), nitric acid, and sodium hydroxide scrubbers to treat off-gases. The biggest cost is for electrical power to operate the electrolysis T-cells.

Evaluation

The CerOx process avoids some, but not all, of the deficiencies of the Ag(II) process. Cerium is much cheaper than silver and much less toxic, and its release to the environment is not as strictly regulated. Unlike the Ag(II) process, the CerOx process could potentially be used to treat both RRS and MMD neutralents, although the high concentration of chlorine in RRS neutralent would result in the formation of large amounts of toxic chlorine gas that would have to be scrubbed. Like the Ag(II) process, CerOx uses large quantities of corrosive nitric acid and generates large quantities of nitrogen oxides at the cathode, which must be reformed and the waste gases scrubbed. Finally, CerOx is not as mature a technology as Ag(II) and would require a larger investment for further development.

The CerOx process was initially developed by the Lawrence Livermore National Laboratory and patented by the Battelle Pacific Northwest Laboratory. Recently, it has been licensed to CerOx Corporation, which has constructed and is operating a small unit at the University of Nevada in Reno for the destruction of laboratory organic wastes. The unit is designed to treat one 35-gallon barrel of waste at a time.

During a visit to the Reno facility, the committee noted that no provisions were being made for the separation of inorganic salts that accumulate in the anolyte solution. When the salts become too concentrated, the solution must be replaced so they do not plug up the T-cell.

Although the process has been demonstrated with a variety of different wastes, the ownership of the unit has still not been transferred to the university; thus the process remains under development and ownership of the vendor. No doubt Ce(IV) can destroy many common organic wastes, such as methylene chloride. However, the process has never been tested with any of the neutralents. Based on the process chemistry of the unit and discussions with the operator, the committee believes the unit at Reno would be adequate in size to treat all of the neutralents from the MMD and RRS.

In summary, the Ce(IV) process has the potential to treat RRS and MMD neutralents with fewer disadvantages than the Ag(II) process. The Ce(IV) process uses a less toxic, cheaper, and not strictly regulated substance (cerium) and operates at low temperature and pressure. The process has some disadvantages. Large amounts of nitric acid (a corrosive) are used, and large amounts of chlorine gas are generated from the process. Therefore, in terms of the pollution-prevention criteria, the Ce(IV) process was rated poor. However, chlorine gases could be scrubbed using pollution-control equipment. The most serious disadvantage is that the technology is not mature enough for immediate use.

The ratings for electrochemical oxidation Ce(IV) are summarized in Tables 4-3a and 4-3b.

TABLE 4-3a Electrochemical Oxidation Ce(IV): Top Priority Criteria

Criterion	Rating
Technical Effectiveness	
Integral part of a coherent system	Fair. Process has not been interfaced with the neutralent process and has not been used to treat neutralent.
Destruction efficiency	Good.
Inherent Safety	
Minimal storage and transportation of hazardous material	Good. Nitric acid is a hazardous material that is transported routinely.
Minimal toxicity and flammability (process materials)	Good. Cerium is much less toxic than silver.
Containable temperature and pressures	Excellent. Moderate temperatures (90°C) and pressures allow reliable containment.
Pollution Prevention	
Minimum toxicity of effluents	Probably good. Carbon is oxidized to carbon dioxide, hydrogen to water, and nitrogen to either nitrogen oxide (scrubbed) or elemental nitrogen. Sulfur is converted to sodium sulfate. Chlorine is converted to Cl_2, which is scrubbed.
Minimal use of processing materials	Fair. Large quantities of nitric acid and Ce(IV) are required.
Solids, liquids, and gaseous wastes	Poor. Large volumes of gaseous effluents will require scrubbing. Large quantities of water remain.
Minimal number of processing steps	Good. Small number of steps.
Temperatures, pressures, corrosion, plugging, and other operating difficulties minimized to prevent unprogrammed shutdowns	Fair. The system must be glass lined. Joints will have leakage and corrosion problems. Corrosion could lead to plugging of the electrochemical cells. The design of the cell is not known at this time.
Permit Status	
Allowed by regulations and capable of meeting schedules imposed by records of decision and treaties	Not known. Probably OK because release of cerium to the environment is not regulated.

Table 4-3b Electrochemical Oxidation Ce(IV) Process: Important Criteria

Criterion	Rating
Robustness	
Stability and continuity of operation	Good.
Cost	
Minimal total costs	Good. Cerium is much cheaper than silver.
Practical Operability	
Minimal training time	Fair. Training time likely to be substantial.
Continuity	
Vendor likely to remain in operation and raw materials likely to remain available	Fair. Technology provider (CerOx Corporation) is a small company that licenses the process.
Space Efficiency	
Minimal weight, area, and volume of operating equipment	Good. System for demonstration, which is modular, was demonstrated at Aberdeen in spring 2000. Data are not yet available.
Materials Efficiency	
Use of internally recyclable materials	Good. Nitric acid is recycled.
Beneficial reuse of wastes	Not applicable.
Use of externally recyclable materials	Not applicable.

BIODEGRADATION

Biodegradation is not a feasible treatment method for RRS neutralents because chloroform, which is present in high concentrations, is highly resistant to oxidation by this method. As a treatment method for MMD neutralents, biodegradation is also doubtful for a number of reasons, as discussed below.

Description

Biotreatment processes use microorganisms to destroy certain organic compounds in dilute aqueous solutions. Aerobic processes result in the partial or total oxidation of neutralent compounds, although the structures of some compounds render them highly resistant. Oxygen is supplied, usually as air that is sparged into the reactor. Nutrients, such as nitrogen in the form of an ammonium salt, and a carbon source, such as dextrose, are often added.

Evaluation

Biodegradation is generally perceived to be natural and, therefore, a publicly acceptable approach to the destruction of wastes. The operating temperature is near ambient, precluding the formation of chlorinated dioxins and furans. Biodegradation is used to treat sewage in many communities, and the safety and reliability of this technology are taken for granted. The U.S. Army is currently testing chemical hydrolysis as a method of destroying chemical agents (HD mustard) at Aberdeen, Maryland, and VX nerve agent at Newport, Indiana. (The results of the tests were not available at the time this report was written.) The hydrolysate resulting from the treatment of HD mustard agent, which contains thiodiglycol and sodium chloride as the primary reaction products, will be treated at Aberdeen using biodegradation technology. The VX hydrolysate, which contains thiol amine and methyl phosphonic acid as the primary reaction products, is not readily amenable to treatment by biodegradation (NRC, 2000b).

A recent evaluation by industry experts indicates that the biodegradability of several compounds present in RRS and MMD neutralents, such as chloroform and hydantoins (major components of RRS waste streams), are extremely resistant to biotreatment (Dekleva and Gannon, 2000). Other compounds that are present at lower concentrations, such as 1,1,2-trichloroethane and 1,1,1,2-tetrachloroethane, are also very resistant to biotreatment. Because the process also requires that microbes be acclimatized for each composition in the waste streams, the quantities of microbes may not be sufficient to accomplish this task. Thus, biotreatment is not a promising treatment for RRS waste streams.

MMD waste streams may be more amenable to biotreatment although they would have to be diluted at least 100-fold with water, and the carbon-nitrogen-phosphorus ratios would have to be adjusted. However, these streams also contain low levels of chloroform, hexachlorobenzene, and hexachlorobutadiene, all of which are known or expected to be resistant to biotreatment. Thus, the treatment of MMD neutralent by biodegradation is not likely.

The ratings for biodegradation are summarized in Tables 4-4a and 4-4b.

TABLE 4-4a Biodegradation: Top Priority Criteria

Criterion	Rating
Technical Effectiveness	
Integral part of a coherent system	Poor. The biological treatment system cannot be easily transported.
Destruction efficiency	Poor. Not usable for RRS neutralent; destruction efficiency rarely exceeds 90 percent.
Inherent Safety	
Minimal storage and transportation of hazardous material	Good. Nutrients and other raw materials for the process are not expected to be hazardous. If the solid products are hazardous, they will be rendered nonhazardous before release from the site.
Minimal toxicity and flammability (process materials)	Good. No highly toxic or flammable materials used.
Containable temperature and pressures	Excellent. Operating temperatures are near ambient, and the pressure is near atmospheric.
Pollution Prevention	
Minimal toxicity of effluents	Fair. A catalytic oxidizer (catox) may be necessary to destroy organic compounds in the large air stream that passes through the bioreactor. Past studies have shown that measurable amounts of chlorodibenzodioxins and chlorodibenzofurans are formed in the catox, if employed for off-gas treatment.
Minimal use of processing materials	Poor. Very large processing material streams, such as carbon source for bioreactor, are necessary.
Solids, liquids, and gaseous wastes	Poor. Large volumes of gases and large volumes of sludge and salts are necessary.
Minimal number of processing steps	Poor. Numerous unit operations are required to remove reaction products from the bioreactor effluent and recycle the water.
Temperatures, pressures, corrosion, plugging, and other operating difficulties minimized to prevent unprogrammed shutdowns	Good. Mild conditions should lead to minimal unprogrammed shutdowns.
Permit Status	
Allowed by regulations and capable of meeting schedules imposed by records of decision and treaties	Fair. If catox system is employed for treatment of off-gas. Excellent otherwise.

TABLE 4-4b Biodegradation: Important Criteria

Criterion	Rating
Robustness Stability and continuity of operation	Fair. Many unresolved issues for biotreatment process.
Cost Minimal total costs	Fair. Capital and operating costs are expected to be high. Transportation and set-up costs will also be high.
Practical Operability Minimal training time	Good. Operations are conventional, and training should be similar to training for typical chemical plants.
Continuity Vendor likely to remain in operation and raw materials likely to remain available	Good. The company that offers the immobilized cell bioreactor technology (the biological technology of most interest) operates the technology commercially and is large and stable. The anticipated nutrients and other raw materials are commodities and are expected to be commonly available.
Space Efficiency Minimum weight, area, and volume of operating equipment	Poor. Equipment, although conventional, may be very large. The weight and volume will be high, and the footprint will be large.
Materials Efficiency Use of internally recyclable materials	Fair. Some reaction liquor can be internally recycled.
Beneficial reuse of wastes	Poor. Metals, if any, could be recycled, but not other residues.
Use of externally recyclable materials	Not applicable.

SOLVATED-ELECTRON TECHNOLOGY

SET would be an inappropriate technology for the treatment of MMD neutralents, which have a high water content. The sodium reagent used for the operation of SET would react with water causing a release of hydrogen gas and requiring excessive quantities of reagent for further treatment. SET could potentially be used to treat RRS neutralents with very little, if any, water.

Description

SET involves the reaction of organic compounds with solutions of metallic sodium in anhydrous liquid ammonia. When sodium is dissolved in liquid ammonia, it forms sodium cations. The electrons released in the process are solvated by ammonia and are highly mobile in the solution. Teledyne-Commodore has proposed using SET for the treatment of assembled chemical weapons and also as a treatment technology for RRS neutralents in response to Stone & Webster's Commerce Business Daily announcement.

Whereas most of the technologies discussed in this report are oxidation processes, SET is a chemical reduction process (the only other reduction process is GPCR, described below).

The SET process can be carried out at −33°C, the boiling point of liquid ammonia, and at atmospheric pressure, or at ambient temperature and slightly elevated pressures (125 psia to 182 psia).

In general, solvated electrons are attracted to the covalent bond between carbon and a more electronegative species, such as chlorine, fluorine, phosphorus, sulfur, or oxygen. The result is a rupture of molecular bonds and some molecular reorganization leading to a complex mixture of chemical species.

After the reduction process has gone to completion, as indicated by persistence of the intense bright blue color of the SET solution, the reaction products are hydrolyzed. Hydrolysis destroys the excess sodium with the release of hydrogen and brings about other reactions that are not yet fully understood.

Evaluation

SET has the advantage of operating at low temperatures and low to moderate pressures, although the engineering advantages may be outweighed by the difficulty of handling anhydrous ammonia and sodium metal, both of which are

TABLE 4-5a Solvated-Electron Technology: Top Priority Criteria

Criterion	Rating
Technical Effectiveness	
Integral part of a coherent system	Fair. Not suitable for MMD neutralent and not tested for RRS neutralent.
Destruction efficiency	Fair. Not usable for aqueous waste stream of the MMD.
Inherent Safety	
Minimal storage and transportation of hazardous material	Fair. Uses liquid ammonia and metallic sodium.
Minimal toxicity and flammability (process materials)	Poor. Ammonia is toxic; both ammonia and metallic sodium have caused fires.
Containable temperature and pressures	Good. Operates at ambient temperature and slightly elevated pressures that are easy to control.
Pollution Prevention	
Minimum toxicity effluents	Fair. Gases released from the SET process and subsequent hydrolysis are mainly hydrogen, ethane, and ethylene. The liquid product contains complex organics, which must be further treated by oxidation. The final oxidation products are suitable for landfill.
Minimal use of processing materials	Fair. Requires metallic sodium and anhydrous liquid ammonia; both present handling challenges.
Solids, liquids, and gaseous wastes	Good. Hold-test-release systems are used for effluent gases, and the likelihood of explosion is very low. Aqueous effluents can be treated with caustic or hypochlorite. Solid wastes can be stabilized in cement and disposed of in a landfill.
Minimal number of processing steps	Fair. Processing steps include SET, hydrolysis, oxidation, solidification/ stabilization, and ammonia recovery.
Temperatures, pressures, corrosion, plugging, and other operating difficulties minimized to prevent unprogrammed shutdowns	Poor. Optimum conditions not well defined; final oxidation step not well understood.
Permit Status	
Allowed by regulations and capable of meeting schedules imposed by records of decision and treaties	Fair. Would require a Subtitle X permit and a great deal more work.

toxic and have been known to cause fires. Based on tests conducted by Teledyne-Commodore, SET is very effective with highly halogenated materials and, therefore, may be able to mineralize the main constituents of RRS neutralents. However, the agent breakdown products are likely to be transformed into residual organic compounds that will require additional treatment (Mitretek, 1999). In addition, SET process efficiency is poor when treating process aqueous waste streams, such as MMD neutralents.

SET is also less mature than some of the other treatment technologies considered by the committee. The optimum operating conditions have not been defined, and the final hydrolysis step is not fully understood. The treatment steps for the vapor streams have not been defined. SET also requires a Subtitle X permit,[5] which would require that uncertainties about its operation be resolved.

The ratings for electron technology are summarized in Tables 4-5a and 4-5b.

---

[5]Both liquid/gaseous ammonia and metallic sodium are widely used in industry, but not for treatment of hazardous wastes. A Subtitle X permit is required under RCRA for processes that treat wastes regulated as hazardous in nonstandard ways. (Standard treatment processes, such as incineration and tank treatment, are permitted under other RCRA subtitles.)

TABLE 4-5b Solvated-Electron Technology: Important Criteria

Criterion	Rating
Robustness Stability and continuity of operation	Poor. Many unresolved issues with respect to the final oxidation step.
Cost Minimal total costs	Fair. Capital and operating costs are likely to be high. Materials cost are relatively low.
Practical Operability Minimal training time	Poor. The process is unusual and, therefore, will require specialized training of personnel.
Continuity Vendor likely to remain in operation and raw materials likely to remain available	Excellent. Teledyne Brown Engineering has been in business since 1953 and is a division of Teledyne Technologies, an $800-million publicly traded company. Commodore Applied Technologies is also publicly traded but is much smaller and less financially stable. Supplies, mainly sodium and ammonia, are readily available commodities.
Space Efficiency Minimal weight, area, and volume of operating equipment	Poor. While the equipment may be transportable, not enough is known to characterize the system as space efficient.
Materials Efficiency Use of internally recyclable materials	Good. Ammonia is recovered.
Beneficial reuse of wastes	Fair. Possibility of recovery of energy from off-gases.
Use of externally recyclable materials	Fair. The ammonia is recovered for recycling.

WET-AIR/O_2 OXIDATION

WAO is a promising treatment for both RRS and MMD neutralents. WAO operates under slightly more aggressive temperature and pressure conditions than chemical oxidation processes. The process is used commercially and has an established track record with compounds similar to those found in neutralents.

Description

The WAO process oxidizes and hydrolyzes organic contaminants in water at temperatures of 150°C to 315°C and pressures of 150 psia to 3150 psia, below the critical temperature of water and pressure (374°C and 3,204 psia). If pure oxygen is used instead of air as the oxidizing agent, the gas volumes that must be managed are greatly reduced.

Organic compounds containing carbon, hydrogen, and oxygen are converted to carbon dioxide, water, and short-chain, biodegradable compounds, such as acetic acid and formaldehyde. Depending on reaction conditions, further biotreatment of residues may be necessary. Toxic heavy metals in the neutralent would have to be precipitated and filtered out prior to biotreatment. Sulfur-containing organics are mineralized to sulfate ions in solution; phosphorus-containing organics are converted to phosphate ions; chlorine-containing organics are converted to chloride ions; nitrogen-containing organics are converted to ammonium ions, nitrate ions, nitrogen gas, or nitrous oxide gas, depending on the organic nitrogen compound; cyanides are converted to carbon dioxide and ammonium ions.

Two U.S.-based vendors, Battelle and Zimpro (now part of U.S. Filter), have demonstrated successful WAO equipment. Battelle's assisted hydrothermal oxidation process uses WAO or SCWO to destroy halogenated and other wastes under conditions that avoid the formation of acid gases. Oxygen or other oxidants are often added. Battelle claims the process thoroughly destroys (mineralizes) organic wastes, including chemical warfare agents, at substantially faster oxidation rates than SCWO at comparable operating temperatures. Reaction times for WAO range from one to two hours. Gases are treated with a thermal oxidizer prior to release, and no provision is made for gas containment. Battelle has used the technology on a small pilot scale in a continuous mode.

Zimpro has installed more than 300 WAO systems worldwide. The process has been used commercially to treat spent

TABLE 4-6a Wet-Air/O_2 Oxidation: Top Priority Criteria

Criterion	Rating
Technical Effectiveness	
Integral part of a coherent system	Good. The RRS and MMD neutralents might have to be diluted with water to a chemical oxygen demand (COD) of 120,000 mg/L or less, but the system can be field-erected or prefabricated in modules.
Destruction efficiency	Fair. Short-chain biodegradable organics may remain in the effluent and require further treatment. Methyl phoshonic acid is resistant and would have to be studied further. Arsenic would be converted to arsenate, which would have to be precipitated and filtered out prior to final polishing with biodegradation.
Inherent Safety	
Minimal storage and transportation of hazardous material	Excellent.
Minimal toxicity and flammability (process materials)	Excellent. No reagents required other than water and air or oxygen.
Containable temperature and pressures	Excellent. High destruction efficiency can be achieved at temperatures under 300°C and pressure under 2,075 psia.
Pollution Prevention	
Minimal toxicity of effluents	Good. Gas phase contains carbon dioxide, oxygen, and nitrogen and no dioxins. Sulfur, phosphorous, and chlorine remain in the liquid phase as dissolved salts, and oxygen probably does as well. The liquid phase also contains biodegradable organics. Solids produced can be stabilized, placed in drums, and disposed of in a permitted landfill.
Minimal use of processing materials	Excellent. Needs water and air or oxygen.
Solids, liquids, and gaseous wastes	Fair. Most wastes are aqueous liquids but are potentially suitable for discharge to a POTW.
Minimal number of processing steps	Good. Process uses pumps, compressors, heat exchangers, and a reactor. Final effluent requires biological treatment.
Temperatures, pressures, corrosion, plugging, and other operating difficulties minimized to prevent unprogrammed shutdowns	Good. Technology provider (Zimpro) has not experienced corrosion, plugging, or other operating difficulties in 300 commercial installations.
Permit Status	
Allowed by regulations and capable of meeting schedules imposed by records of decision and treaties	Excellent. Process has been designated best available data by the EPA for many land-banned hazardous wastewaters.

caustic liquors, high-strength petrochemical wastewater streams, coke oven gas liquors, and municipal sludge from ethylene facilities and petroleum refineries.

Evaluation

WAO is a strong candidate for the treatment of both RRS and MMD neutralents because the process requires only the addition of water and air or oxygen, and no dioxins are formed (in fact, Battelle claims dioxins are destroyed). WAO is most effective on dilute aqueous solutions (e.g., chloroform must be diluted to less than 20,000 mg/L), and RRS and MMD neutralents might have to be diluted with water to reduce their chemical oxygen demand.[6] When a titanium liner is used, no evidence of corrosion has been observed in experimental studies conducted with the feedstocks, temperatures, and pressures described below. This is a major advantage over SCWO technologies.

WAO is currently used in more than 300 commercial installations, and the Environmental Protection Agency (EPA) has specified WAO as a best-demonstrated available technology for the treatment of hazardous wastewater containing

---

[6] For RRS and MMD neutralents, a dilution of 4 to 14 times may be necessary for maximum effectiveness.

TABLE 4-6b Wet-Air/O$_2$ Oxidation: Important Criteria

Criterion	Rating
Robustness	
Stability and continuity of operation	Good. Tests would be necessary on RRS and MMD neutralents specifically, but process has a long commercial history of treating other refractory organics.
Cost	
Minimal total costs	Good. Costs are reported as less than incineration, but the effluent does need to be treated biologically to achieve complete mineralization.
Practical Operability	
Minimal training time	Good. Temperatures and pressures are moderate, and the process is similar to others used in the chemical industry.
Continuity	
Vendor likely to remain in operation and raw materials likely to remain available	Excellent. Zimpro, the principal vendor is part of U.S. Filter, which in turn is owned by Vivendi, the largest environmental firm in the world.
Space Efficiency	
Minimum weight, area, and volume of operating equipment	Good. Skid mounted units are available.
Materials Efficiency	
Use of internally recyclable materials	Poor. None of the materials used are recyclable.
Beneficial reuse of wastes	Poor. Waste is mainly wastewater which could potentially be treated for reuse but currently is not.
Use of externally recyclable materials	Poor. Raw materials are air or oxygen and water, neither of which is available in recycled form.

a variety of wastes classified under P and U in 40 CFR 261. U.S. Filter has tested the destruction of many types of wastes in WAO testing laboratories and pilot-plant facilities. In tests on pesticides, reported destruction efficiencies were greater than 99 percent for malathion at 200°C, dyfonate at 260°C, parathion at 260°C, and glyphosate at 260°C to 280°C. For example, in a test on chloroform reported in 1985, concentration was reduced from 4,500 mg/L to 3 mg/L in 60 minutes at 275°C (Dietrich et al., 1985). The Illinois Waste Management and Research Center at the University of Illinois in Champaign-Urbana also has several reactors designed to carry out WAO studies.

WAO has achieved excellent destruction efficiencies to biodegradable compounds with inorganic and organic cyanides, chlorinated aliphatics, halogenated aromatics containing nonhalogen functional groups, and amines. Halogenated aromatics without other functional groups (e.g., PCBs and chlorobenzene), and alkyl phoshonic acids are relatively resistant. Zimpro has done bench-scale tests on 30 organic compounds, including acenaphthene, acrylonitrile, 2-chlorophenol, 2,4-dinitrotoluene, 1,2-diphenyl hydrazine, pentachlorophenol, chloroform, carbon tetrachloride, 1,2-dichloroethane, hexachlorcyclopentadiene, chlorobenzene, 2,4-dichloroaniline, 2,4,6-trichloroaniline, 1-chloronaphthalene, malathion, Kepone, Arochlor 1254, and 1,2-dichlorobenzene. They achieved greater than 99 percent destruction efficiencies in bench-scale tests of WAO on 2-chlorophenol, chloroform, carbon tetrachloride, 1,2-dichloroethane, 2,4-dichloroaniline, 2,4,6-trichloroanaline, and 1-chloronaphthalene (Dietrich et al., 1985). Excellent destruction efficiencies were obtained for biodegradable compounds with inorganic and organic cyanides, chlorinated aliphatics, halogenated aromatics containing nonhalogen functional groups, and amines. Zimpro successfully treated these organic compounds by conducting these tests at temperatures between 240°C and 280°C and at pressures of 750 psia to 2,075 psia. The committee expects these ranges to be comparable for processing of neutralents in both the MMD and RRS.

WAO operates at somewhat elevated temperatures and pressures (150°C to 315°C and 150 psia to 3,150 psia). WAO may be effective in treating neutralents from RRS and MMD, but some organics that require further treatment may remain in the effluents. Effects of high salt concentration on rates of destruction remain to be tested.

The ratings for WAO are summarized in Tables 4-6a and 4-6b.

SUPERCRITICAL WATER OXIDATION

SCWO has the potential to treat both RRS and MMD neutralent waste streams, producing a relatively clean effluent containing mainly inorganic salts and water. However, high operating temperatures and pressures, as well as corrosion and potential plugging caused by salts, have made it difficult to operate SCWO processes on a routine basis.

Description

SCWO is a hydrothermal process for the oxidative destruction of organic wastes. The system involves introducing air or oxygen in the presence of high concentrations of water heated above the critical temperature and pressure of pure water (374°C, 3,204 psia).[7] The reaction mechanisms for the destruction of organic compounds generally involve free-radical chain reactions with oxidative radicals. Thermal bond cleavage and polar or ionic reactions, including hydrolysis, also occur.

In the SCWO process, the feed stream (aqueous waste) to the reactor is heated, pressurized, mixed with oxidant, and pumped through a flow reactor at supercritical conditions designed to provide the required residence time (typically, several seconds to a minute). Heat produced by the reaction can be recovered (or must be removed) based on the heating value of the feed stream. If the heating value is too low to heat the reactor, supplemental fuel can be added. Downstream of the reactor, the system is depressurized, either before or after cooling. Solids produced from oxidation reactions can be recovered prior to or following pressure letdown. The effluent is then passed through gas/liquid separators, and the gas stream and aqueous streams can be treated further.

SCWO process effluents include vent gases, liquid effluents (neutralized with NaOH), and crystallized salts, all of which must comply with regulatory requirements prior to disposal. The gases primarily consist of oxygen, nitrogen, and carbon dioxide but may also contain trace quantities (7 to 28 ppb in testing to date) of volatile organic compounds (VOCs). Any remaining VOCs in the effluent gases are filtered out by passing them through activated carbon filters.

Evaluation

In the committee's judgment, SCWO would effectively mineralize agent neutralents. RRS neutralents, however, contain considerable quantities (more than 50 percent) of chloroform and other chlorinated compounds that may be highly corrosive to the SCWO liner. Experience in processing nonstockpile neutralents in a SCWO unit will be necessary to quantify the nature, locations, and rates of corrosion.

Problems associated with the stability of SCWO (e.g., maintenance of temperature and pressure for at least 20 hours, control of salt accumulation) appear to have been resolved for a test of SCWO processing of NaOH-based VX neutralents. In testing on materials of construction with this neutralent, salts accumulated at a rate of about one pound per hour, limiting runs with neutralent to 20 hours (runs with surrogate continued for about 40 hours) (Dekleva and Gannon, 2000). However, issues related to the mechanisms and locations of salt buildup, the chemical composition of the salts produced, and the effectiveness of flushing away salts are still unresolved. Pressure containment is another issue that must be addressed.[8]

A removable titanium or platinum liner for processing nonstockpile neutralents has been suggested by the SCWO technology proponent because it could withstand reactions with acidic chloride compounds found in RRS neutralents. Although the corrosion of reactor liners and the erosion of metal parts may be mitigated by periodic replacement of the liner and frequent maintenance, the associated frequency, cost, and down time will have to be determined.

The ratings for SCWO are summarized in Tables 4-7a and 4-7b.

---

[7] See NRC, 1998b, Table 3-1, for comparisons of operating SCWO systems.

[8] "Failure of the pressure containment system (piping, SCWO reactor, post-reactor air cooler, or pressure let-down system) could result in rapid depressurization and dispersal of hot fluids and debris at high velocities. Similarly, failure of the pressure let-down system could result in a large pressure surge that could rupture equipment downstream. ... The pressure let-down system may be the weak link in the full-scale SCWO process chain" (NRC, 1998b).

TABLE 4-7a Supercritical Water Oxidation: Top Priority Criteria

Criterion	Rating
Technical Effectiveness	
Integral part of a coherent system	Fair. Oxidation of neutralents from the RRS and MMD can be the final step following neutralization of nonstockpile chemical agent fill.
Destruction efficiency	Good. Neutralization of agent and RRS or MMD must reduce agent concentration in neutralent to under 50 ppm as per RCRA permit. SCWO should reduce any remaining agent to below detection limits.
Inherent Safety	
Minimal storage and transportation of hazardous material	Fair. Neutralents are under engineering controls. Liquid effluent is evaporated, and salts are crystalized, placed into drums, and sent to a hazardous waste landfill following TCLP testing.
Minimal toxicity and flammability (process materials)	Good. Some kerosene is used to start the process.
Containable temperature and pressures	Poor. Operates at high temperature and pressure, 650°C and 3,400 psia. Pressure containment failure can result in ejection of liquid, but volume is limited and pressure relief valves prevent over-pressurization. Pressure containment failures, resulting from overtemperature (reducing the strength of the containment material) has occurred. Water pump failure is a typical problem.
Pollution Prevention	
Minimal toxicity of effluents	Good. Liquid effluent is evaporated, and salts are crystallized, put into drums, and sent to a hazardous waste landfill following TCLP testing. Gases (oxygen, nitrogen, and carbon dioxide) are filtered and released. Brines and salts are tested and put into drums and can be sent to hazardous waste landfill. Fate of arsenic compounds is an issue.
Minimal use of processing materials	Good. Requires start-up fuel, oxidant (air), and NaOH to reduce acidity of brine.
Solids, liquids, and gaseous wastes	Fair. Most wastes are liquids that can be evaporated to obtain leachable salts, which can be disposed of in a Subtitle C landfill.
Minimal number of processing steps	Good. Uses a reactor, compressors, pressure let-down, evaporators, effluent air coolers, pumps, and gas/liquid separator.
Temperatures, pressures, corrosion, plugging, and other operating difficulties minimized to prevent unprogrammed shutdowns	Poor. Corrosion and plugging have been serious problems, but pilot SCWO at Dugway Proving Ground has operated for several 20-hour runs using VX hydrolysate without plugging and with stable temperature and pressure. Extent of unprogrammed shutdowns is not known because of lack of operating experience with nonstockpile neutralents for representative times.
Permit Status	
Allowed by regulations and capable of meeting schedules imposed by records of decision and treaties	Fair. The contractor expects RCRA Part B permit for SCWO at Newport Chemical Agent Disposal Facility from the state regulatory agency.

TABLE 4-7b Supercritical Water Oxidation: Important Criteria

Criterion	Rating
Robustness Stability and continuity of operation	Fair. Using NaOH-based VX hydrolysates, 20-hour runs at stable temperatures and pressures have been achieved. No operating experience with nonstockpile neutralents.
Cost Minimal total costs	Good. Capital and operating costs are expected to be moderate.
Practical Operability Minimal training time	Fair. Need more information on training time and necessary skill levels. Maintenance time requirements may be high if frequent shutdowns for salt removal are necessary. Process monitoring and control strategies should to be tailored to SCWO.
Continuity Vendor likely to remain in operation and raw materials likely to remain available	Good. Technology provider has operating experience and will be operating a unit processing VX hydrolysates at a chemical stockpile disposal facility in Newport, Indiana. Vendor also operates a unit in Utah.
Space Efficiency Minimum weight, area, and volume of operating equipment	Fair. The reactor at Dugway Proving Ground, which has a volume of 1.51 ft^3 and, at its targeted flow rate of 17.7 lbs/hr, can process 141.6 lbs of material per 8-hour day. This is less than half of the reactor's design flow rate, and, thus, the Dugway reactor should be able to process about 300 lbs per 8-hour day. Four or five flatbed trucks are required to transport the unit and its ancillary equipment.
Materials Efficiency Use of internally recyclable materials	Excellent. Evaporator condensate is recycled. Otherwise, there is nothing to recycle.
Beneficial reuse of wastes	Poor. Solid wastes are put into drums and sent to a hazardous waste landfill. No beneficial uses for this material.
Use of externally recyclable materials	Not applicable.

GAS-PHASE CHEMICAL REDUCTION

Although GPCR appears to be capable of destroying nonstockpile neutralents from both the RRS and MMD, the process generates large volumes of effluent gases that require a complex treatment system.

Description

The GPCR process uses hydrogen and steam at temperatures of approximately 850°C at atmospheric pressure to convert organic wastes into substances that are either less toxic or convertible to less toxic materials. The overall process requires a high-temperature reaction vessel followed by a gas-scrubbing train to remove inorganic by-products. Residence time in the reactor is only a few seconds.

In the GPCR reactor, which contains a hydrogen-rich atmosphere, organic chemicals are reduced to methane, water, carbon soot, and other by-products, including acid gases, such as HCL from chloroform in RRS neutralent, hydrogen sulfide, phosphorus-containing products from VX neutralent, and arsenic-containing products from lewisite neutralent. These products, as well as the carbon soot, must be treated or scrubbed, adding to the complexity of the process. Catalytic steam reformers supply hydrogen gas to the GPCR reactor by steam reforming of natural gas. Vertical radiant tube heaters with internal electric heating elements are used to heat the inside of the reactor. When the gases leave the reactor, they pass through primary and secondary caustic scrubbers to remove acid gases, water, and fine particulates. The gas stream exiting the secondary scrubber is a mixture of hydrogen, methane, carbon monoxide, carbon dioxide, nitrogen, and trace quantities of light hydrocarbons. This gas is stored in tanks, sampled, and tested. If permitted, the gas can be used as an auxiliary fuel for the steam boiler.

Evaluation

GPCR, a well established thermal treatment technology operating in a pyrolytic mode, is capable of very high destruction efficiency. GPCR is the only one of the eight processes considered that involves the chemical reduction of neutralent with hydrogen and steam. The process generates large volumes of effluent gases compared to low-temperature oxidation processes. GPCR is a complex process that requires the management of hot hydrogen gas in the reactor, separate scrubbing of effluent acid gases, the recovery of phosphorus and arsenic, and the control of carbon soot buildup in the reactor.

The technology provider has stated that GPCR reactors have been operating reliably in Canada and Australia for several years at commercial scales (several tons per day) to treat a variety of feedstocks, including polychlorinated biphenyls (PCBs), polychlorinated aromatic hydrocarbons, dichlorodiphenyltrichloroethane, chlorobenzenes, and toluene. According to the process developer, the reactor can operate effectively over a temperature range of 750°C to 950°C.

Testing with HD, VX, and VX hydrolysates has been conducted in laboratory and bench-scale tests at the Edgewood Research, Development and Engineering Center. In 1996, 1,440 grams of VX and 2,450 grams of HD were destroyed with a destruction efficiency of 99.999999 percent (eight 9's) in a portable pilot reactor. Tests with nonstockpile neutralents, however, have not been conducted.

To the best of the committee's knowledge, no commercial scale GPCR reactor has been permitted to operate in the United States. As a result, processing nonstockpile neutralents by GPCR may be delayed because no regulatory experience with either the process or with destruction of the feedstock for this process (MMD and RRS neutralents) is available. In addition, the proposed use of the GPCR off-gas as a boiler fuel poses unique permitting challenges (NRC, 1999b). A demonstration of the GPCR system was completed in September 2000 as part of the second set of ACWA demonstrations (Demo II). Results were not available at the time this report was written.

The ratings for GPCR are summarized in Tables 4-8a and 4-8b.

TABLE 4-8a Gas-Phase Chemical Reduction: Top Priority Criteria

Criterion	Rating
Technical Effectiveness	
Integral part of a coherent system	Good. Part of technology package to dispose of stockpile chemical munitions that can be a stand-alone process for nonstockpile CWM; can process either neat agent or neutralent.
Destruction efficiency	Good. High efficiencies have been achieved with similar compounds.
Inherent Safety	
Minimal storage and transportation of hazardous material	Good. Hydrogen is transported and stored routinely.
Minimal toxicity and flammability (process materials)	Fair. Prevention of leaks important to contain hydrogen.
Containable temperature and pressures	Fair. Reactor operates at near ambient pressure but at 850°C.
Pollution Prevention	
Minimal toxicity of effluents	Good. Scrubbed gases containing hydrogen, methane, CO, and CO_2. Solids containing phosphorous, arsenic, and possibly sulphur must be disposed of.
Minimal use of processing materials	Excellent. Process requires only fuel, NaOH for caustic scrubbing, and water.
Solids, liquids, and gaseous wastes	Poor. Primary emissions are gases, some of which are recycled as boiler fuel. Carbon soot is also produced and must be disposed of.
Minimal number of processing steps	Fair. Process is moderately complex, especially for scrubbing reactor effluent gases. Process involves feed to the reactor, the reactor itself, scrubbing of gases, steam reforming, and handling of materials and storage equipment. Because process is integrated, it must be controlled to allow excess materials to be recirculated.
Temperatures, pressures, corrosion, plugging, and other operating difficulties minimized to prevent unprogrammed shutdowns	Fair. Potential for plugging as a result of carbon (soot) buildup. Reactor must be cleaned periodically to prevent this. Operating experience with neutralents containing chlorine, phosphorous, and other nonstockpile constituents will show if plugging and corrosion are problems.
Permit Status	
Allowed by regulations and capable of meeting schedules imposed by records of decision and treaties	Undetermined. Full-scale reactor has not been permitted in the United States (but was tested as part of ACWA Program). Data is not yet available. Trial burns to obtain RCRA and other permits may be required. Permit for using reactor off-gas as a boiler fuel will be required.

TABLE 4-8b Gas-Phase Chemical Reduction: Important Criteria

Criterion	Rating
Robustness	
Stability and continuity of operation	Excellent. A reactor operating since 1995 in Australia is capable of processing liquid feedstocks, including PCBs, polyaromatic hydrocarbons, chlorobenzenes, and dichlorophenyltrichloroethane. Start-up and shutdown have been demonstrated. More information necessary about scale-down of reactor to meet nonstockpile needs and about operation with neutralents.
Cost	
Minimal total costs	Good. Appears to be competitive, but no cost data are available for processing small quantities of neutralents.
Practical Operability	
Minimal training time	Good. Appears to be well established for materials being processed but not for operation with neutralents. Additional scrubbers, permitting concerns, materials handling requirements (e.g., arsenic, phosphorous, carbon) may reduce practical operability.
Continuity	
Vendor likely to remain in operation and raw materials likely to remain available	Excellent. Vendor operates facilities in Canada and Australia and will be part of ACWA Demo II. Expected to remain in business. No unique raw materials required.
Space Efficiency	
Minimal weight, area, and volume of operating equipment	Good. Appears to be scalable. Not all size requirements for equipment, including scrubbers and steam reformers, are known.
Materials Efficiency	
Use of internally recyclable materials	Good. Some of the reactor effluent gas is recyclable as boiler fuel.
Beneficial reuse of wastes	No. Liquid and solid wastes are not recyclable but must be disposed of.
Use of externally recyclable materials	Not applicable. Facilities (e.g., landfills) that will accept these wastes have not been identified.

PLASMA-ARC TECHNOLOGY

Plasma-arc technology is a very high-temperature process that probably would effectively destroy both RRS and MMD neutralents although the large quantities of water present in MMD neutralents could present problems. Plasma-arc generates large volumes of high-temperature vapor streams that require containment and high-quality treatment.

Description

Plasma-arc technologies utilize electrical discharges in various gases to produce a field of intense radiant energy and high-temperature ions and electrons that cause dissociation of chemical compounds in a containment chamber. The process, operating at temperatures as high as 20,000°C, occurs in a closed hearth reactor. The reaction chamber is maintained at slightly less than atmospheric pressure to prevent the release of hazardous effluents. Material exposed to the plasma environment is transformed into atoms, ions, and electrons. However, small molecules form as the gases leave the reaction zone.

Three types of waste streams are produced: plasma off-gas that is first treated to completely destroy VOCs and then cleaned by a two-stage scrubber, followed by a sophisticated filtration system prior to release; wastewater from the water treatment system used in the purification of the off-gases; and vitrified inorganic products that fall to the bottom of the containment vessel.

Evaluation

Plasma-arc systems can achieve high destruction efficiencies, reportedly higher than 99.99999 percent (seven 9's). They are most efficient when used to treat low-volume, highly concentrated feed streams. Consequently, they would be less efficient when used with neutralents from the RRS and MMD, the latter of which would require that the system process large amounts of water. The committee is concerned

TABLE 4-9a Plasma-Arc Technology: Top Priority Criteria

Criterion	Rating
Technical Effectiveness	
Integral part of a coherent system	Excellent. Could be integrated with existing technologies. System available in either a fixed or mobile configuration.
Destruction efficiency	Excellent. Has achieved destruction efficiencies of seven 9's or better with similar chemical compounds in Germany and Switzerland.
Inherent Safety	
Minimal storage and transportation of hazardous material	Excellent. Requires no storage of hazardous materials.
Minimal toxicity and flammability (process materials)	Excellent. Requires no storage of toxic or flammable materials.
Containable temperature and pressures	Poor. Temperatures can run as high as 20,000°C. Subatmospheric pressures permit reliable containment.
Pollution Prevention	
Minimum toxicity of effluents	Good. Exhaust gases are filtered and scrubbed. Arsenic is removed from the scrubber, recycled, and eventually forms a slag.
Minimal use of processing materials	Fair. Helium or another suitable gas used in starting the process is replaced by nitrogen or air. Water is used as a coolant.
Solids, liquids, and gaseous wastes	Poor. Converts the neutralent primarily into an off-gas that requires extensive treatment. Solids are formed into a slag.
Minimal number of processing steps	Good. Requires a pollution control system, but there are not many steps.
Temperatures, pressures, corrosion, plugging, and other operating difficulties minimized to prevent unprogrammed shutdowns	Good. System requires only weekly preventive maintenance based on an 80-hour-per-week operating schedule. Torch electrode is replaced after 20 hours.
Permit Status	
Allowed by regulations and capable of meeting schedules imposed by records of decision and treaties	Poor. System has not been permitted to operate in the United States. May be considered by regulators as incineration.

about the very high operating temperatures of this system and the need for extensive off-gas treatment.

A plasma-arc process was tested for the destruction of assembled chemical weapons for the Army's ACWA Program. The test configuration included a 300-kW unit that used nitrogen as the plasma gas. Tests conducted on dimethyl methyl phosphonate and hydrolysates of HD and VX achieved high destruction efficiencies but generated products of environmental concern, including C_2N_2, hydrogen cyanide, and metal cyanides. The by-products with a different plasma gas would be different.

The Committee on Review and Evaluation of Alternative Technologies for Demilitarization of Assembled Chemical Weapons concluded, in concurrence with the Army and the Dialogue (a citizen group), that the plasma torch apparatus demonstrated for the ACWA Program did not qualify for further consideration for the demilitarization of assembled chemical weapons. That committee noted, however, that "the variety of equipment problems encountered in the demonstration were due to the immaturity of the particular demonstration equipment and not due to a fundamental inability of plasma-based technologies to achieve acceptable results" (NRC, 1999b).

A patented plasma-arc process, PLASMOX®, is currently operational in Europe in both fixed and mobile configurations. The process combines features of pure plasma and incineration operations. The first step in the process is a plasma treatment unit, which is started on helium but is maintained during processing by using either nitrogen or air as the plasma gas. Operation with nitrogen would represent a pure plasma mode, while operation on air would represent a plasma-augmented incineration mode. Off-gases from the plasma unit are fed to a secondary combustion chamber employing air or oxygen at 1,000°C to 1,200°C. This operation is a main feature of incinerators. The off-gas handling system includes a gas cooler and quencher, a two-stage

Erratum

The wording of a finding (page 4, first finding, left column; repeated on page 52, third finding, left column) as printed in the final published National Research Council report *Disposal of Neutralent Wastes* (ISBN 0-309-07287-5, published in March 2001) contains an error. The words "over incineration" were erroneously added during the publication process and should be deleted. The committee did not compare incineration to the alternative technologies assessed in the report. The NRC-approved finding is as follows:

Finding. The committee identified some low-temperature, low-pressure, less complex technologies that might be used to treat neutralent waste. The benefits of these technologies include low worker risk, public acceptance, low risk to the surrounding community, and simplicity of operation.

TABLE 4-9b Plasma-Arc Technology: Important Criteria

Criterion	Rating
Robustness	
Stability and continuity of operation	Excellent.
Cost	
Minimal total costs	Good. Capital and operating costs are expected to be average. Costs of transportation and setup will be low.
Practical Operability	
Minimal training time	Good. Operations are simple, training in the handling of toxic chemicals should be similar to training for typical chemical plants.
Continuity	
Vendor likely to remain in operation and raw materials likely to remain available	Good. Large, stable engineering company. The equipment is comprised mostly of off-the-shelf items.
Space Efficiency	
Minimum weight, area, and volume of operating equipment	Good. Mobile system can be moved on three standard trailers.
Materials Efficiency	
Use of internally recyclable materials	Good. Majority of waste products recycled internally.
Beneficial reuse of wastes	Fair. Metals can be recycled, but not other residues.
Use of externally recyclable materials	Not applicable

scrubber, a dust filter, a high-efficiency particulate air filter, and an activated carbon filter. When treating chlorinated materials, the process generates small quantities of dioxins, which are removed in the carbon filter. A mature technology that has been operated for extended periods of time in Europe, this technology has been used to destroy lewisite, mustard, adamsite, and phosgene.

The ratings for plasma-arc technology are summarized in Tables 4-9a and 4-9b.

OVERALL RANKINGS

The committee's qualitative demarcation of the axes in Figure 4-1 into low, moderate, high, and very high temperature and pressure ranges corresponds roughly with the extent of engineering controls required for safe operation. Biodegradation, chemical oxidation, and electrochemical oxidation fall into the low-temperature, low-pressure range; SET, when operated at room temperature, operates at slightly higher pressures. WAO is operated over a range of moderate temperatures and moderate to very high pressures. GPCR and plasma-arc processes (as well as the Army's baseline incineration technology) operate in high and very high-temperature, low- to moderate-pressure regimes; SCWO requires high temperatures and very high pressures.

Other things being equal, the lower the operating temperature and pressure required for a technology, assuming that it can achieve acceptable performance, the more likely it is to be simpler, cheaper, and less risky to health and the environment. As the operating temperature and pressure increase, several undesirable changes may occur: the sophistication (and cost) of the required engineering controls increases; undesirable by-products may be formed (e.g., tar soot, polyaromatic hydrocarbons, chlorodibenzodioxins); and containment and the treatment of high-temperature vapor streams become issues of concern. SCWO might be an exception to the general rule that undesirable by-products may be formed in high-temperature processes. The committee began with a bias toward technologies in the low-temperature, low-pressure category. Nevertheless, after much study and debate, two of these (electrochemical oxidation and SET) were given lower rankings because of other concerns (e.g., toxicity and the corrosiveness of process chemicals). Biodegradation was the only technology that was judged to be unacceptable.

The technologies are described below in order of their ranking, from highest to lowest.

Low-Temperature Chemical Oxidation in Water

Chemical oxidation in water at less than 100°C appears to be the most attractive technology. This technology has a reasonable chance of destroying MMD and RRS neutralent successfully and does not require exotic or significantly toxic

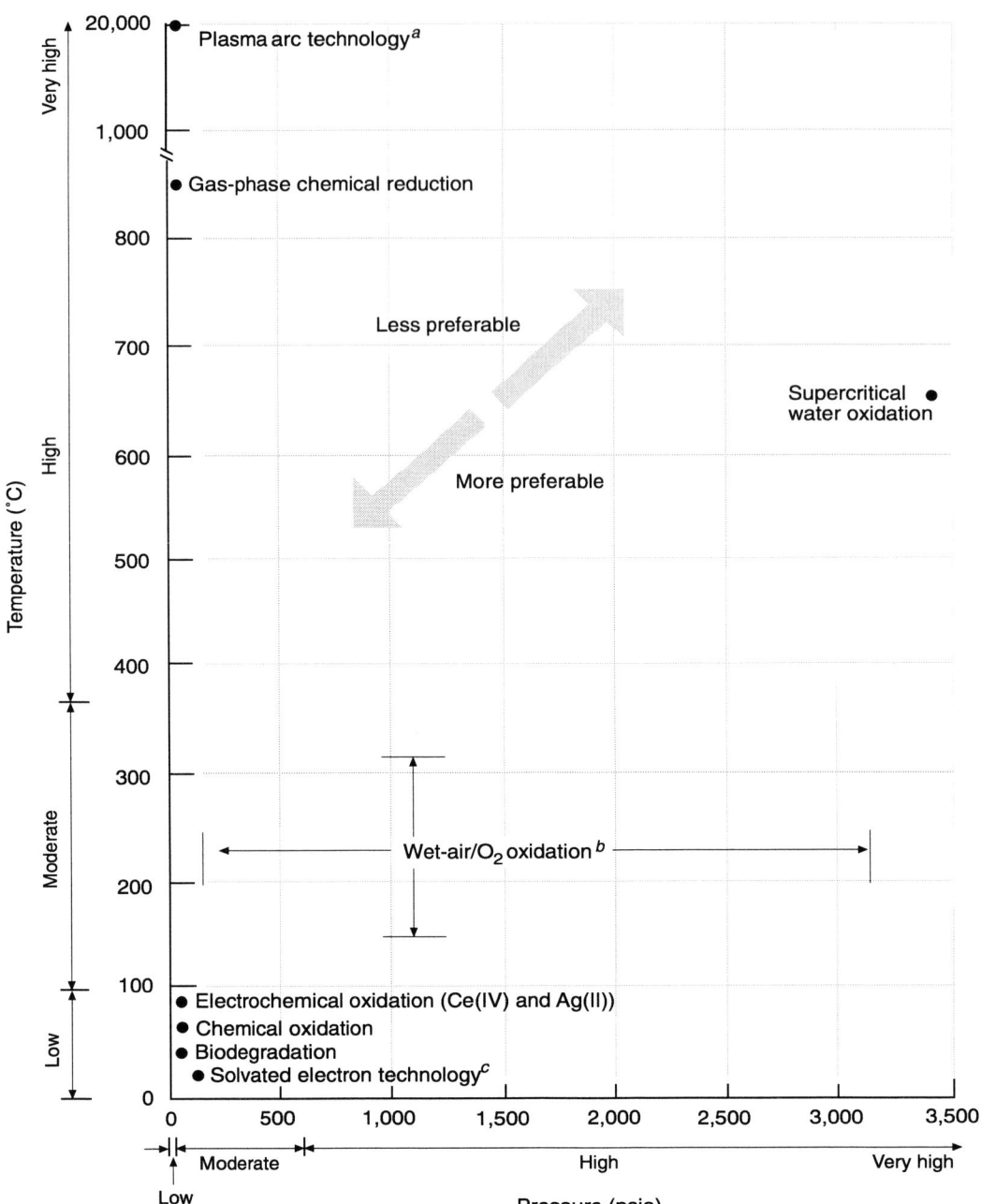

FIGURE 4-1 Comparative operating temperatures and pressures.

chemicals. Other than carbon dioxide from the oxidation of carbon-containing molecules, the process should not produce a vapor stream. Control of acidity by adjusting the pH will be necessary when handling chlorinated materials. Problem metals, such as arsenic and mercury, would be captured by demonstrated technologies, such as ion-exchange, precipitation, or activated carbon, and disposed of through recycling or in a RCRA landfill. Teledyne-Commodore, an existing vendor, has proposed using peroxidisulfate (oxidant species) to treat MMD neutralents.

Wet-Air/O_2 Oxidation

WAO has almost all of the advantages of chemical oxidation but two significant disadvantages: (1) the process usually uses an air stream as the source of oxygen; and (2) WAO operates at higher temperature and pressure than chemical oxidation. The air stream could contain oxides of nitrogen, which must be treated before discharge, although one vendor (Zimpro) has not found detectable levels of nitrogen oxides in the effluent from any of the tests conducted with air. Use of pure oxygen instead of air does create safety problems. Therefore, the concentrations of organics in the effluent gas must not exceed the lower explosive limits of oxygen. Substituting oxygen for air would significantly reduce the amount of vapor that would have to be contained and cleaned. A process using oxygen instead of air would be a higher temperature analog of the chemical oxidation technologies described above. Zimpro is already set up to demonstrate this technology and has already demonstrated it using oxygen in the treatment of other organic compounds.

Oxidation in Water by Cerium IV

Although oxidation in water by Ce(IV) is a low-temperature process, it may produce a significant vapor stream because of the formation of chlorine. The process also uses nitric acid, which must be regenerated. Oxidation by Ce(IV) should be pursued only if low-temperature chemical oxidation and WAO cannot break down the neutralents to the point that the resultant materials could be discharged to a POTW or equivalent facility. *However*, because there is an existing vendor for this technology, and if actual or synthetic neutralent can be made readily available, it may be possible to test this technology with a minimum of time and effort. If so, this should be done as a potential backup technology, although low-temperature chemical oxidation should remain the technology of choice.

Oxidation in Water by Silver II

The only advantage of oxidation in water by Ag(II) over oxidation by Ce(IV) is that Ag(II) may have greater oxidizing power. Otherwise, this technology has several disadvantages. Silver salts are far more toxic and expensive than cerium salts. In addition, silver forms a highly insoluble, toxic salt with chloride, which could be present in large amounts and would require extensive handling, collection, and recycling. Ag(II) technology should be considered a low priority for development funding. *However*, because an existing vendor for this technology is already conducting tests under the ACWA Program, if actual or synthetic neutralent can be made readily available, it may be possible to test this technology with a minimum of time and effort. If so, that should be done as a potential backup, although low-temperature chemical oxidation should remain the technology of choice.

Supercritical Water Oxidation

SCWO is a potentially simpler process than any of the technologies discussed so far *except* for low-temperature chemical oxidation, the primary choice. However, developers have been trying unsuccessfully to solve the corrosion and plugging problems associated with SCWO for almost two decades. The Army is attempting to demonstrate an improved SCWO process for the Chemical Stockpile Disposal Program and for the ACWA Program. Unless the Army demonstration is very successful, however, SCWO should only be considered for treatment of RRS and MMD neutralents if there is reasonable assurance that further investments will not be prohibitive. *However*, if actual or synthetic neutralent can be made readily available, it may be possible to test this technology with a minimum of time and effort in combination with the Army's tests of SCWO for the Chemical Stockpile Disposal Program. If so, that should be done. SCWO could be a potential backup technology but should not compromise the preference for low-temperature chemical oxidation.

Solvated-Electron Technology

SET uses metallic sodium in anhydrous liquid ammonia, both of which require careful and complicated handling. In addition, liquid ammonia requires refrigeration or operation at moderate pressure. This technology has no obvious advantages over the preferred technologies for destruction of RRS neutralents and is not suitable for treatment of MMD neutralents. SET should, therefore, be ranked lower than low-temperature, low-pressure, less-complex process technologies.

Gas-Phase Chemical Reduction

GPCR technology would generate a voluminous, high-temperature vapor stream that would require significant efforts to contain, scrub, and otherwise treat. The technology has apparently been demonstrated by vendors outside the United States to the satisfaction of the regulators and the community. Except for stronger destruction capability,

GPCR has no obvious advantages over the preferred technologies for destruction of RRS and MMD neutralents and should be ranked lower than low-temperature, low-pressure, less-complex process technologies.

Plasma-Arc Technology

Like GPCR, plasma-arc technology would generate a large, very high temperature vapor stream that would require significant efforts to contain, scrub, and treat. This technology has apparently been demonstrated by vendors outside the United States to the satisfaction of the regulators and the community of the hosting country. Other than stronger destruction capability, plasma-arc technology has no obvious advantages over the preferred technologies and should be ranked lower than low-temperature, low-pressure, less-complex process systems.

Biodegradation

The committee could not find any record of tests of stand-alone biological processing for treating any form of RRS or MMD neutralent. (However, an immobilized-cell biological reactor was tested by Parsons/AlliedSignal and was found to be effective for treating mustard hydrolysate,[9] a different type of waste stream). Aeration of the bioponds might generate significant secondary air emissions that would require treating large volumes of off-gas. Preliminary modeling suggests that the chloroform in the RRS neutralent would be particularly difficult to destroy by biodegradation. Small amounts of metals and biologically recalcitrant organic compounds in both the RRS and the MMD neutralents may also pose major problems. Thus, stand-alone biological processing would probably be applicable only to MMD neutralent, if at all. In addition, a test of using biodegradation to treat GB and VX hydrolysate was unsuccessful, which raises questions about the effectiveness of using biodegradation to treat GB and VX neutralent. Overall the committee felt that no investment in biodegradation was warranted for destruction of neutralents.

---

[9]The Army refers to the destruction of chemical agent via hydrolysis as chemical neutralization. The term is derived from the military definition of "neutralize," to render something unusable or nonfunctional. Hydrolysis is a reaction of a target compound with water, often catalyzed by an acid or a base, in which a chemical bond is broken in the target, and the components of water, OH^- and H^+, are inserted at the site of the bond cleavage. Hydrolysate is the product resulting from hydrolysis. The technical definition of neutralization is a chemical reaction between an acid and a base to form a salt and water. Neutralent is the product generated from neutralization.

5

Public Acceptance and Regulatory Considerations

In 1985, after determining that incineration was a safe, proven technology, the Army decided to use incineration technology to destroy stockpile CWM (PMCD, 1999). Evaluations by the Army and others continue to show that the incineration of stockpile chemicals can be performed without threatening human health or the environment (NRC, 1997; PMCD, 1999). However, in 1994, in response to public concerns about incineration and at the direction of Congress, the Army began evaluating alternatives to incineration (PMCD, 1999). Public concerns about incinerating wastes may also affect the disposal of nonstockpile chemicals and neutralents. Therefore, the NSCMP has also found it necessary to evaluate alternative, nonincineration technologies.

Two constraints on the selection of a disposal technology are regulatory requirements and public stakeholder involvement in the selection and use of that technology. These two influences on the decision-making process are interrelated. Thus, the public may be not only directly involved in the Army's decision making but may also indirectly affect that decision by its involvement in the determining of regulatory requirements. (The Army's successful disposal of nonstockpile materiel within the specified deadlines may depend on whether the Army successfully addresses the public involvement and regulatory issues.)

The Statement of Task governing the actions of this committee does not explicitly request an assessment of the efficacy of the Army's nonstockpile public involvement program. The primary task before this committee was to evaluate the technical merits of alternative treatment technologies. However, after hearing the views of many stakeholders, and based on the Army's experience with chemical demilitarization programs, the committee concluded that the public policy framework for selecting a technology must be taken into consideration and that public acceptability should be a criterion for selecting a technology.

PUBLIC ACCEPTANCE

Public acceptance is the result of a *process* that involves identifying interested or affected stakeholders, clarifying issues, and putting in place mechanisms to facilitate reaching an agreement (NRC, 1996a).[1] In the context of the NSCMP, stakeholders include: the government agency making the decision (i.e., the Army); Congress, which enacted the statutes requiring that the Army make a decision; local citizens who may be affected by the decision; national nonprofit groups involved in the public policy debate; contractors, who must implement decisions; and federal and state regulatory agencies.

Through previous research and case studies, and through the stockpile program's recent experiences with the ACWA Program (described below), productive ways of resolving contentious issues over technology selection have been identified, the value of public involvement in governmental decisions in general has been documented, and the issues of concern to the public in the selection of disposal technologies for chemical materiel have been clarified. The necessity, as well as the desirability, of proactively seeking public involvement in policy decisions that once were considered purely scientific has been well documented (e.g., Funtowicz and Ravetz, 1985; Walsh, 1990; Wynne, 1996). The following discussion is a summary of these findings, as well as the recent experiences of the Chemical Stockpile Disposal Program and NSCMP.

Incineration has been the subject of citizen opposition for many years and the target of numerous lawsuits that have

---

[1]This report can only touch on the extensive literature on the balance between scientific analysis and public deliberation and the wide variety of formal and informal mechanisms for facilitating discussion (NRC, 1996b). For a practitioner's guide to public involvement, see Creighton, 1999; for a more theoretical approach, see Renn et al., 1995.

focused media attention on the issue and led to a change in government policy. The problem of politically stigmatized wastes (e.g., napalm, and now, stockpile and nonstockpile wastes) has become so common that the EPA has dubbed them "wastes of concern" (i.e., wastes that are likely to create significant public concerns) (EPA, 2000a). Because of public opposition, the Army was directed by Congress (P.L. 102-84) to evaluate alternative disposal options (i.e., technologies that might be significantly safer and more cost effective than the baseline incineration system).

During the deliberations for this study, the committee reviewed previous studies to identify reported public views of disposal technologies (see especially NRC, 1996a, 1999b), and monitored the Nonstockpile Chemical Weapons Citizens Coalition (NSCWCC) and Chemical Weapons Working Group websites and publications that highlight public views of the nonstockpile program. In addition, the committee solicited the views of two stakeholder groups—opponents of incineration and federal regulators. Although these "representatives" do not reflect the full spectrum of public opinion, both groups are expected to be active participants in the decision to develop and deploy nonincineration technologies. The committee gathered this information to get a sense of what these two groups considered important, both for determining public acceptability and for gaining a general understanding of regulatory problems that might arise.

The committee solicited the views of citizen groups in several ways. NSCWCC was asked to provide documents outlining its views on nonincineration technologies.[2] Two committee members separately observed two meetings of the CORE group (Army personnel from the chemical demilitarization program, representatives of regulatory agencies, and representatives of citizen groups) that were scheduled during the study period. The CORE group, which meets once or twice a year to discuss public issues, provided an opportunity for committee members to observe interactions between participants and the NSCMP, to hear directly their issues of concern, and to talk informally with members of the group. In addition, representatives of these groups availed themselves of an opportunity to attend open committee sessions and present their views before the entire committee. One citizen representative of the CORE group accompanied the committee on a site visit to observe a technology being evaluated. Several members of the committee met with a group of federal regulators from the EPA Office of Waste Programs.

In both formal and informal discussions with members of the committee, representatives of citizens groups affirmed their strong opposition to the use of incineration for the primary or secondary treatment of CWM (NSCWCC, 2000), underscored their commitment to the development and deployment of an alternative technology (or technologies), expressed their belief in the long-term storage of neutralent wastes if an alternative is not available in the near term. The concerns expressed by these representatives are summarized below.

The "dialogue process" established by the ACWA Program was cited in the briefings as a model for early, direct public involvement in technology decisions. Through the ACWA Dialogue process, representatives of diverse public groups (including citizens and regulators) participated in the early stages of decision making. In other words, the public was involved in establishing criteria for selecting and demonstrating technologies, as well as in making trade-offs. The goal of the ACWA Dialogue process was to incorporate public concerns and preferences before a policy was set and to diminish or avoid the conflicts, delays, and cost escalations incurred earlier by the Stockpile Program. In the opinion of the National Research Council committee that evaluated the alternative technologies for the ACWA Program, the dialogue process was "a positive step toward gaining acceptance for alternative disposal technologies" (NRC, 1999b). In the opinion of the participants, one of the most important results of the dialogue process was that it engendered trust (NSCWCC, 2000).

The NSCMP has already established a mechanism for public involvement based on the ACWA experience. To date, the program has convened three meetings of a group of representatives of public interest groups, regulators, and NSCMP personnel to facilitate interactions and discussions. Although the group is still recruiting representatives with diverse public views and developing effective working relationships, its establishment is an important initial step in seeking public input and improving working relations between NSCMP program staff and the public.

Some representatives of citizens groups who briefed the committee felt strongly that the Army should consider storing neutralent until an alternative technology to incineration could be developed and permitted. The Army's opposition to the storage of neutralent is based on several factors. As discussed in Chapter 2, MMD neutralent can contain breakdown products that are on the list of Schedule 2 precursors (chemicals that could be used to remanufacture chemical agents) under the CWC. The CWC requires that Schedule 2 precursors derived from existing agents be destroyed in the same time frame as the chemical agents. The Army is concerned that long-term storage may be considered failure to

---

[2]Documents provided included comments on the following documents: (1) Nonstockpile Draft Programmatic Environmental Impact Statement; (2) RRS System Test Plan; (3) Pre-Operational Surveys for the RRS and MMD-1; (4) Draft RCRA Part B Permit, Subpart X for the MMD; and (5) Draft RCRA Part B Permit, Subpart X for the RRS (NSCWCC, 2000). Additional reports included *Technical Criteria for the Destruction of Stockpiled Persistent Organic Pollutants* (Greenpeace, 1998), and *The American People's Dioxin Report* (Center for Health, Environment and Justice, 1999).

comply with the CWC or may be misinterpreted as a precedent for another party to the treaty to store large quantities of precursors that could be quickly converted into chemical agents. In addition, long-term storage would be inconsistent with the regulatory requirements under RCRA that limit the length of time hazardous wastes can be stored to 90 days. The Army's long-term interest, as well as the public's, would be served by a description in writing of the Army's position. This document could then be the basis for discussion on the limitations on the Army's ability to store neutralent solutions.

Public opposition to incineration includes the perceived instability of the process, the potential for explosion, and the potential for unplanned releases of harmful pollutants. The formation and dispersal of dioxins and furans are of particular concern, as well as the release of minimal amounts of unknown, but potentially high-risk chemicals. To lessen these concerns, it was suggested by NSCWCC that the NSCMP evaluate incineration technology against the same criteria the ACWA Program has adopted for evaluating nonincineration technologies (NSCWCC, 2000).

According to the spokespersons for the NSCWCC who briefed this committee, the active opposition of some public interest groups is based not only on technical issues but also on other concerns, such as a desire for environmental justice, the unfairness of the decision-making process, mistrust of the technology provider and the Army, and the lack of accountability and institutional safeguards (e.g., environmental monitoring and emergency preparedness) (NSCWCC, 2000).

Previous research and case studies have shown that the active opposition of some public interest groups to the Army's baseline technology suggest that the Army's goal should be not only compliance with federal or state regulations, but also the highest possible performance standards and protection of workers and public health. In this context, briefers listed some of the criteria by which citizens evaluate technologies (Bradbury et al., 1994). They emphasized, however, that the criteria are not rigidly fixed. An alternative technology that does not meet all of the criteria might still be acceptable to a community—and community preferences may vary. Their primary criteria were:

- containment of by-products and effluents for analysis and further processing, if necessary
- identification of all by-products and effluents
- low-temperature, low-pressure operation
- no dioxins or furans
- pollution prevention (i.e., generation of as little secondary waste as possible)

In discussions with the committee and in a subsequent written statement, the representatives described the broad context for citizens' evaluations of technologies. First, they asserted that the destruction rate efficiency of a technology is not an accurate measure of its net environmental and public health impact. Attributes of preferred technologies include production of low volumes of hazardous waste and high destruction efficiency, rather than simply dilution, of hazardous waste; waste should not simply be moved from one place to another. However the waste is destroyed, it should directly impact as few communities as possible, and, as much as possible, the by-products should be reprocessed or recycled.

REGULATORY STAKEHOLDERS

In the broadest sense, federal and state regulatory authorities can be considered stakeholders. However, regulatory agencies have the legal authority to bar some options from consideration and/or require the implementing agency to take specific actions to comply with environmental, health, safety, and treaty requirements.

The interaction between citizen stakeholders and regulatory stakeholders is unique. Since the 1960s, environmental laws have created a role for the public in the development of regulatory requirements, particularly if they relate to human health and safety. Thus, virtually every federal and state statute requires public notice and public comment. Some statutes, such as Superfund, require that regulatory agencies consider community opinion (42 U.S.C. § 9621, Section 21).

Environmental Protection Agency

Several committee members met with a group of federal regulators from the EPA Office of Waste Programs to discuss the regulatory and permitting challenges to the development of new nonincineration technologies. In addition, information and perspectives were obtained from state regulators. The following subsections summarize these discussions but do not endorse a particular option; the committee had neither a mandate nor the resources to explore the advantages and disadvantages of each option. The Army will have to coordinate with the primary permitting authorities (i.e., the states), the EPA, and the public to develop a regulatory permitting plan. Some or all of the alternatives discussed below may be found to be disadvantageous, infeasible, or not cost effective. However, the process of working with stakeholders earlier, rather than later, to determine the necessary information and requirements will be advantageous to the Army and will ultimately speed up the process.

Neutralents as Hazardous Waste

Neutralents may be designated hazardous wastes pursuant to EPA rules. However, states may interpret these rules differently. In the committee's opinion, neutralents will be handled, transported, disposed of, or treated in compliance with regulations for hazardous wastes or hazardous materials. Neutralents that are designated hazardous wastes may not be stored at a site for more than 90 days without a RCRA permit, unless a specific exception applies (EPA, 2000b; NRC, 1999a). Any new nonincineration disposal technology must be permitted, unless it falls under a provision that allows treatment without a permit. For example, under EPA guidelines, treatment in a tank within 90 days does not require a hazardous waste permit (EPA, 2000a; Weddle, 1993; Williams, 1987a, 1987b). However, because states issue federal permits, and because they may have different interpretations of EPA regulations, they may require a permit for this treatment option. If treatment is performed pursuant to a Superfund cleanup, then no federal or state hazardous waste treatment permit is required for any portion of the treatment (Section 124(e) of CERCLA, 42 U.S.C. § 9224(e)).

Lack of Specific Requirements for Alternative Technologies

No EPA regulations or guidances specify the levels of destruction, levels of air emissions, water discharge limits, or other environmental requirements for alternative treatment technologies used to treat nonstockpile materiel. Although some established environmental requirements may be applicable, EPA and the states usually develop environmental requirements on a case-by-case basis.

Delays in Implementation

For the past four years, the Army has been going through the process of obtaining an environmental permit (RCRA part B) for the treatment of CAIS and other nonstockpile materiel in the RRS and MMD. Based on this experience and experiences with other nonstockpile and stockpile chemical treatment systems (i.e., incineration), the regulators expressed concerns (which are shared by the committee) that it may be very difficult to complete the regulatory process for a new technology in the time frame of the NSCMP/CWC. To expedite the process, the NSCMP will have to develop a regulatory compliance plan to evaluate its options in cooperation with EPA, state environmental regulatory agencies, and DOT.

EPA regulators expressed concerns that waiting until an alternative technology has been selected would unduly delay the process and expressed a willingness to begin working immediately with the Army, states, and interested public. In the committee's opinion, the Army should invite stakeholders to participate in an environmental-criteria working group to develop regulatory requirements and begin the process of determining which regulatory requirements apply to the alternative technologies being considered. The following options could be considered.

Treatment in a Tank

EPA rules allow nonthermal treatment of hazardous wastes in a tank on site without a hazardous waste permit, as long as the treatment is completed in 90 days or less (EPA, 2000b; Weddle, 1993; Williams, 1987a, 1987b). Under these conditions, neither RRS nor MMD neutralents would require a federal hazardous waste permit. However, state environmental regulatory agencies could impose more stringent permit requirements that would preclude this option. Nevertheless, this approach could potentially save the Army time and expense. Whether this exemption applies to a particular treatment operation must be determined on a case-by-case basis based on federal rules and guidance, as well as state rules and guidance.

National Guidance

EPA drafts national model permit guidance typically on an industry-by-industry basis. The Utah permit for testing the RRS and MMD could become the basis for a model permit for treating neutralents. The Army could work with EPA, states, and local citizens to ensure that the treatment meets the regulatory requirements of this policy and addresses other applicable environmental requirements. The development of environmental requirements by this approach would be the same as for permit-by-rule. The end point, however, would be nonbinding guidance, rather than a legally binding rule.

Exemptions, Exceptions, and Variances

Some exemptions, exceptions, or variances from the hazardous wastes rules may apply to neutralents if nonstockpile CWM is remediated under the Superfund rules before being treated in the RRS and MMD. For example, the handling of wastes is exempt from the normal RCRA permitting requirements if the waste is being cleaned up pursuant to the Superfund on-site rule, which preempts all federal and state procedural permitting requirements for hazardous substances treated on a Superfund site (Section 120(e) of Superfund, 42 U.S.C.§9620(e)). Thus, the Army might try to clean up CAIS and nonstockpile chemicals under the Superfund rules.

The on-site exemption from permitting, however, still requires that the Superfund remedy meet the substantive environmental and health requirements of RCRA regulations, unless a Superfund waiver applies. In any case, the proposed cleanup would be subject to public comment under the Superfund public involvement procedures. Many cleanups by the U.S. Department of Energy are currently implemented as Superfund cleanups to obtain this flexibility (EPA, 2000b). At RCRA-regulated facilities, EPA and the state may decide to proceed pursuant to a state-issued federal RCRA permit or a CERCLA action.

Statutory Changes

A narrow statutory amendment could be devised to clarify which requirements apply to the treatment of nonstockpile chemicals and neutralents from the RRS and MMD. For example, Congress could include an exemption similar to the one in Superfund legislation in the nonstockpile legislation. Waivers could include a requirement for state and public input and compliance with the substantive requirements of federal and state environmental requirements. A statutory amendment should only be pursued after consultation with and involvement of the entire stakeholder community to reach a political consensus.

Regulatory Approaches

All of the approaches suggested above would provide flexibility in the development of the regulatory requirements for neutralization of nonstockpile chemicals and the treatment of resulting neutralents. Each approach would allow public stakeholders to participate fully, would allow different options to be pursued, and would expedite the process of implementing an alternative treatment technology.

State Hazardous Waste Permitting Process

The states are the lead agencies in most hazardous waste permitting. Therefore, the Army must involve state regulatory stakeholders (as well as public stakeholders). Given the complexity of the issues and the limits on state budgets, state permitting would be expedited if the Army, EPA, and state regulatory officials developed a model state permit and generic risk assessments for the treatment of nonstockpile neutralent.

6

Findings and Recommendations

Based on the preceding evaluation, site visits, discussion with stakeholders, and information gathered from presentations and other sources, the committee developed a number of findings and recommendations.

TECHNICAL ISSUES

Finding. The committee did not find any experimental studies on the destruction of neutralent wastes generated by the RRS or MMD. Therefore, the analyses of candidate technologies are based on their demonstrated performance with chemically similar materials, as well as on fundamental principles of chemistry and chemical engineering.

Finding. Based on the amount of neutralent expected from planned operations at Deseret Chemical Depot and Dugway Proving Ground, the volume of neutralents generated by the RRS and MMD is expected to be relatively small—on the order of 5,000 gallons per year in normal operation. As a point of reference, a standard tanker truck contains 5,000 to 10,000 gallons, and a railcar may contain as much as 30,000 gallons. Because the facility for disposing of neutralent will not have to handle large volumes or have a high throughput, it could be a laboratory or pilot-plant-scale facility. Thus equipment for technologies currently under investigation for stockpile CWM might be used cost effectively for treating nonstockpile neutralents. At this small scale, all of the technologies reviewed by the committee could be adapted to "semi-fixed, skid-mounted" configurations (see Statement of Task).

Finding. The committee identified some low-temperature, low-pressure, less complex technologies that might be used to treat neutralent waste. The benefits of these technologies over incineration include low worker risk, public acceptance, low risk to the surrounding community, and simplicity of operation.

Finding. The Army's evaluation of alternative technologies must meet the time constraints of the CWC, which requires that all nonstockpile CWM in storage at the time the convention was ratified be destroyed by 2007. Thus far, no alternative incineration technologies have been tested on real, or even simulated, nonstockpile neutralent generated by either the RRS or the MMD. Therefore, bench testing and scale-up demonstrations of candidate technologies with neutralents will be necessary. Because testing the effectiveness of alternatives and determining regulatory limits will take time, the Army may have to fall back on its current incineration strategy for the destruction of neutralent, which includes the use of commercial incinerators, or even the use of the Army's stockpile incinerators.

Finding. Some of the candidate alternatives to incineration for destroying MMD and RRS neutralents involve hardware that has already been developed, and using them would simply require substituting neutralent for existing feeds. For example, one or more of the demonstration units tested for the chemical disposal programs (e.g., ACWA Program) might be used. Because the volume of nonstockpile neutralents will be small, even if the technology is not rated highly according to the committee's criteria but is inherently safe, the savings in time and development costs might justify consideration of this alternative. Demonstration units could be used at their present sites or moved, either as needed or to a mutually agreeable location based on a plan developed with the affected communities and regulatory authorities.

Recommendation. The Non-Stockpile Chemical Materiel Program should pursue a two-track strategy similar to the one adopted by the committee during its selection of a technology: (1) an evaluation of the potential of Assembled Chemical Weapons Assessment demonstration technologies and mature commercial technologies; and (2) technologies that would require further development and investment.

Recommendation. As part of the track-one strategy, the Army should take advantage of available equipment that would require little or no investment (i.e., either alternative technologies from the Assembled Chemical Weapons Assessment [ACWA] Program or existing commercial technologies, such as chemical oxidation, wet-air/O_2 oxidation, or PLASMOX®). The following technologies from the ACWA demonstrations should be considered: electrochemical oxidation Ag(II), gas-phase chemical reduction, solvated-electron technology, and supercritical-water oxidation. If any of these can accomplish the task safely, it might provide a relatively rapid and inexpensive course of action.

Recommendation. If Assembled Chemical Weapons Assessment (ACWA) or the commercial technologies require substantial modifications to processes or permits, the Army should focus first on the most easily adaptable commercial technologies (i.e., chemical oxidation and wet-air/O_2 oxidation). Only if these technologies prove to be unsuitable should the Army consider investing resources in the further development of ACWA technologies (listed below in order of preference):

- electrochemical oxidation with Ag(II) and Ce(IV)[1]
- supercritical-water oxidation
- solvated-electron technology
- gas-phase chemical reduction
- plasma-arc technology

Recommendation. The Army should not invest in further development of biodegradation, which was judged least likely to be effective.

REGULATORY ISSUES AND PUBLIC INVOLVEMENT

The recent experience of federal agencies has shown that the involvement of diverse public groups (including state and federal regulators) is crucial to timely decision making. Stakeholder involvement is particularly important for decisions involving analytical, engineering, or other scientific uncertainties about the protection of human health and the environment. The Army's implementation of an alternative technology or technologies to incineration could be delayed unless regulatory requirements have been developed and the public has been involved in the decision-making and selection process.

The NSCMP could improve its existing public involvement program by (1) exploring ways to ensure representation of diverse public groups in assessments of disposal technologies and associated regulatory issues; and by (2) working closely with potential host communities to identify and address their concerns.

A comprehensive regulatory compliance plan that involves all stakeholders could be essential to the timely implementation of an alternative technology. An environmental criteria working group, with representatives of the Army, EPA regulators, state regulators, officials of the U.S. Department Health and Human Services, public interest groups, and citizens at large, could be formed to undertake advanced planning with the goals of (1) ensuring that substantive regulatory requirements can be met and (2) determining if additional testing or evaluations will be necessary to satisfy public or regulatory concerns.

Finding. Citizens groups that met with the committee strongly urged that the Army consider the long-term storage (i.e., longer than one year) of neutralents rather than incineration. Storage, they argued, would ensure that the Army would have sufficient time to develop, test, and obtain regulatory approval of alternatives to incineration. The committee believes that the Army's mission could be affected by the manner in which it responds to these public concerns.

Finding. The Army provided several reasons for not storing neutralent. First, storage might make it impossible to meet the treaty deadlines for the destruction of the nonstockpile chemical weapons. Second, the Army might be required to meet rigorous, long-term environmental requirements. Third, long-term storage would be inconsistent with regulatory requirements limiting storage time for hazardous wastes. Finally, the cost of storage might be disproportionately high.

Recommendation. To solicit public understanding, and perhaps acceptance, in its decision on whether or not to store neutralent, the Army should issue a detailed white paper explaining the legal, scientific, regulatory, and institutional issues involved. The paper should explicitly describe how risk to the public and workers would be affected by the long-term storage of neutralent prior to its disposal.

Finding. The committee's discussions with citizen groups indicated a need for, and the value of, public involvement in the Army's decisions concerning the selection, deployment, and employment of technologies for disposing of nonstockpile chemical materials.

Recommendation. The committee recommends that the Army expand its public involvement program regarding disposal of nonstockpile chemical materiel. Enough time should be scheduled and enough resources allocated to ensure that the decision-making process is open and that members of the public are involved in determining trade-offs related to the selection, siting, deployment, and employment of disposal technologies.

---

[1] Although not an ACWA technology, this variant of electrochemical oxidation, Ce(IV), should be evaluated.

References

Battelle. 1997. Evaluation of the Vesicating Properties of Neutralized Sulfur Mustard. Final Report. Columbus, Ohio: Battelle.

Bradbury, J.A., K.M. Branch, J.H. Heerwagen, and E.B. Liebow. 1994. Community Viewpoints of the Chemical Stockpile Disposal Program. Washington, D.C.: Battelle, Pacific Northwest National Laboratory.

Brankowitz, W. 2000. U.S. Army Non-Stockpile Chemical Materiel Product (NSCMP) Project Overview/Status. Presentation by William Brankowitz, Office of the Project Manager, Non-Stockpile Chemical Materiel, to the Committee on Review and Evaluation of the Army Non-Stockpile Chemical Materiel Disposal Program, National Research Council, Washington, D.C., May 8, 2000.

Center for Health, Environment, and Justice. 1999. The American People's Dioxin Report. Falls Church, Va.: Center for Health, Environment, and Justice.

Cooper, J.F., G.B. Balazs, and P. Lewis. 1999. Direct Chemical Oxidation: An Application to Hazardous Waste Treatment in Demilitarization. Final Report to Joint DOE/DOD Project, DOD Office of Munitions, FY 1997-8, UCID-ID-134365. May 1999. Livermore, California. Lawrence Livermore National Laboratories.

Creighton, J.L. 1999. How to Design a Public Participation Program. Washington, D.C.: U.S. Department of Energy, Office of Intergovernmental and Public Accountability.

Dekleva, L.A., and J. Gannon. 2000. DuPont's P&B Screening Tool for Prioritization of Chemicals for Risk Assessment. Presented at Environmental Science in the 21st Century: Paradigms, Opportunities, and Challenges. 21st Annual Meeting of the Society of Environmental Toxicology and Chemistry (SETAC) North America, Nashville, Tennessee, November 12–16, 2000.

Dietrich, M.J., T.L. Randall, and P.J. Canney. 1985. Wet air oxidation of hazardous organics in wastewater. Environmental Progress 4(3): 171–177.

DOT (U.S. Department of Transportation). 1997. Dermal Toxicity Evaluation of Neutralized CAIS Nonstockpile Chemical Materiel Program Report. Washington, D.C.: U.S. Department of Transportation.

Dugway Proving Ground. 1998. RCRA permit for MMD-1 testing issued by the State of Utah. Available on line at: http://www.deq.state.ut.us/eqshw/cds/MMDPermit.htm

DuPont Specialty Chemicals. 1992. Data Sheet on Oxone(R) Monopersulfate Compound. Available on line at URL: http://www.dupont.com/oxone/techinfo/index.html (October 2, 2000).

EPA. 2000a. Discussion between members of the Committee on Review of the Non-Stockpile Chemical Materiel Disposal Program and representatives of the EPA Office of Solid Waste, Arlington, Virginia, April 28, 2000.

EPA. 2000b. EPA Toxic Release Inventory, Envirofacts Warehouse. Available on line at URL: http://oaspub.epa.gov/enviro/tris_control.tris_print?tris_id=84074MXMGNROWLE

Funtowicz, S.O., and J.R. Ravetz. 1985. Three Types of Risk Assessment: A Methodological Analysis. Pp. 831–848 in Environmental Impact Assessment, Technology Assessment, and Risk Analysis, edited by V.T. Covello, J.L. Mumpower, P.J. Stallen, and V.R. Uppuluri. New York: Springer-Verlag.

Gieseking, J.K. 1999. Non-Stockpile Waste Streams/Inventory/Monitoring. Briefing by John K. Gieseking, Project Officer for the Office of the Product Manager for Non-Stockpile Chemical Materiel, to the Committee on Review and Evaluation of the Army Non-Stockpile Chemical Materiel Disposal Program, National Research Council, Washington, D.C., June 15, 1999.

Gieseking, J.K. 2000. Personal communication between John K. Gieseking, RRS Program Manager for the Office of the Product Manager for Non-Stockpile Chemical Materiel, and Mr. Greg Eyring, Consultant for the National Research Council. September 2000.

Greenpeace. 1998. Technical Criteria for the Destruction of Stockpiled Persistant Organic Pollutants. Washington, D.C.: Greenpeace.

GRI (Gas Research Institute). 2000. Chemical Oxidation with or without UV Enhancement: Technology Description. Available on line at URL: www.gri.org.

Holm, F.W. 1998. Appendix: Descriptions of Alternative Demilitarization Technologies and Estimated Mass Balances. Pp. 176–177 in Effluents from Alternative Demilitarization Technologies. Vol. 22, NATO ASI Series. Boston, Mass.: Kluwer Academic Publishers.

Hovanek, J.W., L.L. Szafraniec, J.M. Albizo, W.T. Beaudry, V.D. Henderson, Y.-C. Yang, B.K. MacIver, and L. Procell. 1993. Evaluation of Standard and Alternative Methods for the Decontamination of VX and HD in Chemical Agent Disposal Facilities. ERDEC-TR-054. Aberdeen Proving Ground, Md.: Edgewood Research, Development and Engineering Center.

Lawrence Livermore National Laboratory. 2000. Direct Chemical Oxidation of Organic Wastes. Available on line at URL: www-ep.es.llnl.gov

Mikolajczyk, M. 1996. Fundamental Chemistry of Chemical Warfare Agents and Interrelationships in Technologies. Pp. 10–11 in Scientific Advances in Alternative Demilitarization Technologies. Vol. 6, NATO ASI Series. Boston, Mass.: Kluwer Academic Publishers.

Mitretek. 1999. Assessment of ACWA Technologies and Equipment for Treatment of Non-Stockpile Wastes and Chemical Materiel. Technical Report MTR 1999-32V1. McLean, Va.: Mitretek.

Morgan, E.W., R.A. Renne, B. McVeety, R. Johnson, R.L. Phelps, F-S. Yin, J.T. Pierce, J. Blessing, P.W. Mellick, E.J. Olajos, and H. Salem. 1997. Head-Only Inhalation Toxicity Study (LC50) of Chemical Agent Identification Sets (CAIS) Red Process Waste Streams in Rats.

REFERENCES

ERDEC-CR-239. Aberdeen Proving Ground, Md.: Edgewood Research Development and Engineering Center.

Mulholland, K.L., and J.A. Dyer. 1999. Pollution Prevention: Methodology, Technologies, and Practices. New York. American Institute of Chemical Engineers.

NRC (National Research Council). 1994. Recommendations for the Disposal of Chemical Agents and Munitions. Committee on Review and Evaluation of the Army Chemical Stockpile Disposal Program, Board on Army Science and Technology. Washington, D.C.: National Academy Press.

NRC. 1996a. Review and Evaluation of Alternative Chemical Disposal Technologies. Panel on Review and Evaluation of Alternative Disposal Technologies, Board on Army Science and Technology. Washington, D.C.: National Academy Press.

NRC. 1996b. Understanding Risk: Informing Decisions in a Democratic Society. Committee on Risk Characterization, National Research Council. Washington, D.C.: National Academy Press.

NRC. 1997. Risk Assessment and Management at Deseret Chemical Depot and the Tooele Chemical Agent Disposal Facility. Committee on Review and Evaluation of the Army Chemical Stockpile Disposal Program, Board on Army Science and Technology. Washington, D.C.: National Academy Press.

NRC. 1998a. Health Effects of Waste Incineration. Committee on Toxicology, Board on Environmental Studies and Toxicology. Washington, D.C.: National Academy Press.

NRC. 1998b. Using Supercritical Water Oxidation to Treat Hydrolysate from VX Neutralization. Committee on Review and Evaluation of the Army Chemical Stockpile Disposal Program, Board on Army Science and Technology. Washington, D.C.: National Academy Press.

NRC. 1999a. Disposal of Chemical Agent Identification Sets. Committee on Review and Evaluation of the Army Non-Stockpile Chemical Materiel Disposal Program, Board on Army Science and Technology. Washington, D.C.: National Academy Press.

NRC. 1999b. Review and Evaluation of Alternative Technologies for Demilitarization of Assembled Chemical Weapons. Committee on Review and Evaluation of Alternative Technologies for Demilitarization of Assembled Chemical Weapons, Board on Army Science and Technology. Washington, D.C.: National Academy Press.

NRC. 2000a. Evaluation of Demonstration Test Results of Alternative Technologies for Demilitarization of Assembled Chemical Weapons: A Supplemental Review. Committee on Review and Evaluation of Alternative Technologies for Demilitarization of Assembled Chemical Weapons, Board on Army Science and Technology. Washington, D.C.: National Academy Press.

NRC. 2000b. Integrated Design of Alternative Technologies for Bulk-Only Chemical Agent Disposal Facilities. Committee on Review and Evaluation of the Army Chemical Stockpile Disposal Program, Board on Army Science and Technology. Washington, D.C.: National Academy Press.

NSCWCC (Non-Stockpile Chemical Weapons Citizens Coalition). 2000. Assessment of Alternative Technologies by the Non-Stockpile Chemical Weapons Citizens Coalition Working Group. Briefing by an NSCWCC panel to the Committee on Review and Evaluation of the Army Non-Stockpile Chemical Materiel Disposal Program, National Research Council, Washington, D.C., February 23, 2000.

Olajos, E.J., K.P. Cameron, R.A. Way, J.H. Manthei, J.H. Heitkamp, D.M. Bona, and S.A. Thomson. 1996. Acute Dermal Toxicity Evaluation of Product Solutions Resulting from the Chemical Neutralization of HD, GB, and VX via Monoethanolamine (MEA). ERDEC-TR-434. Aberdeen Proving Ground, Md.: Edgewood Research Development and Engineering Center.

Olajos, E.J., H. Salem, and J.K. Gieseking. 1997. Department of Transportation, Dermal Toxicity Evaluation of Product Solutions Resulting from the Chemical Neutralization of HD, GB, and VX via Monoethanolamine (MEA). ERDEC-SP-002. Aberdeen Proving Ground, Md.: Edgewood Research Development and Engineering Center.

PMCD (Program Manager for Chemical Disposal). 1999. Chemical Stockpile Program, Questions and Answers. Available on line at URL: *www.pmcd.apgea.army.mil/text/CSDP/IP/brochures/Q&A/index.html.*

Renn, O., T. Webler, and P. Wiedemann. 1995. Fairness and Competence in Citizen Participation: Evaluating Models for Environmental Discourse. Dordrecht, The Netherlands: Kluwer Academic Publishers.

Soilleaux, R. 1998. Hydrolysis and Oxidation Process Effluents of Some Chemical Warfare Agents and Possible Secondary Treatments. Pp. 29–31 in Effluents from Alternative Demilitarization Technologies. Vol. 22, NATO ASI Series. Boston, Mass.: Kluwer Academic Publishers.

Stone & Webster. 2000. Evaluation of Neutralent Post-Treatment Technologies for the Non-Stockpile Chemical Materiel Program. Prepared by the Technology Evaluation Panel for the U.S. Army Program Manager for Chemical Demilitarization. September 18, 2000. Boston, MA: Stone & Webster Engineering Corporation.

U.S. Army. 1994a. U.S. Army's Alternative Demilitarization Technology Report to Congress. 11 April 1994. Aberdeen Proving Ground, Md.: U.S. Army Program Manager for Chemical Demilitarization.

U.S. Army. 1994b. Detailed Program Plan and Methodology for Developing Alternative Technologies for Chemical Demilitarization. 26 April 1995. Aberdeen Proving Ground, Md.: Alternative Technology Branch, U.S. Army.

U.S. Army. 1996. Survey and Analysis Report, 2nd ed. Aberdeen, Md. Project Manager for Non-Stockpile Chemical Materiel.

U.S. Army. 1999a. Draft Programmatic Environmental Impact Statement. Vol. 1. Transportable Treatment Systems for Non-Stockpile Chemical Warfare Materiel. Aberdeen Proving Ground, Md.: U.S. Army.

U.S. Army. 1999b. Draft Programmatic Environmental Impact Statement. Vol. 2. Appendices: Transportable Treatment Systems for Non-Stockpile Chemical Warfare Materiel. Abderdeen Proving Ground, Md.: U.S. Army.

U.S. Army Research Office. 1994. Destruction of Military Toxic Waste. Presentation at the NATO Advanced Workshop, Naaldwijk, The Netherlands, May 22–27, 1994. Available on line at URL: *www.aro.ncren.net.*

Walsh, W.J. 1990. Making Science, Policy, and Public Perception Compatible: A Legal/Policy Summary, or Do We Want to Clean Up Hazardous Sites or Just Scream and Yell at Each Other. Pp. 206–249 in Ground Water and Soil Contamination Remediation: Toward Compatible Science, Policy, and Public Perception. Report on a Colloquium Sponsored by the Water, Science and Technology Board. Washington, D.C.: National Academy Press.

Weddle, B. 1993. Letter from Bruce Weddle, Acting Director, EPA Office of Solid Waste, to Ethan Ware, Ogletree, Deakins, Nash, Smoak, & Stewart, November 1, 1993.

Williams, M. 1987a. Letter from Marcia Williams, Director, EPA Office of Solid Waste, to Bernard Cox, Hazardous Waste Branch, Alabama DEM, July 1, 1987. Available on line at URL: *http://yosemite.epa.gov/osw/rcra.nsf*

Williams, M. 1987b. Letter from Marcia Williams, Director, EPA Office of Solid Waste, to K. Allford, NL Industries, Inc., March 25, 1987. Available on line at URL: *http://yosemite.epa.gov/osw/rcra.nsf*

Wynne, B. 1996. May the Sheep Safely Graze? A Reflexive View of the Expert-Lay Knowledge Divide. Pp. 44–83 in Risk, Environment, and Modernity: Towards a New Ecology, edited by S. Lash, B. Szerszynski, and B. Wynne. London: Sage Publications Ltd.

Yang, Y-C. 1995. Chemical reactions for neutralizing chemical warfare agents. Chemistry and Industry 9: 334–337.

Yang, Y-C. 1999. Chemical detoxification of nerve agent VX. Accounts of Chemical Research 32(2): 109–115.

Appendixes

A

Biographical Sketches of Committee Members

John B. Carberry (chair), is director of environmental technology for E.I. duPont de Nemours and Company, where he has been employed since 1965; he is responsible for providing technical analysis of existing and emerging environmental issues. Since 1988, he has been involved with initiatives to advance DuPont's environmental excellence through changes in products, recycling of materials, and renewal of processes with an emphasis on reducing waste and promoting affordable, publicly acceptable technologies for the abatement, treatment, and remediation of environmental pollution. Mr. Carberry is chairman of the Chemical Engineering Advisory Board at Cornell University, a fellow of the American Institute of Chemical Engineers, and a member of the Radioactive Waste Retrieval Technology Review Group for the U.S. Department of Energy. He was a member of the National Academy of Engineering (NAE) Committee on Industrial Environmental Performance Metrics. He holds an M.S. in chemical engineering from Cornell University and an M.B.A. from the University of Delaware.

John C. Allen is vice president of transportation at Battelle Memorial Institute. Prior to joining Battelle in 1983, he was a transportation economist and policy analyst with the U.S. Department of Transportation Office of Hazardous Materiel Transportation. Mr. Allen has managed and participated in numerous studies involving the policy, regulatory, institutional, and safety aspects of transporting hazardous and nuclear materials. He has served on various National Research Council (NRC) advisory panels and has been chairman of the Transportation Research Board's Committee on Hazardous Materials Transportation for the past four years. He holds an M.B.A. in transportation from the University of Oregon and a B.A. in economics from Western Maryland College.

Richard J. Ayen, a member of the NRC Committee on Review and Evaluation of Alternative Technologies for Demilitarization of Assembled Chemical Weapons (I and II), received his Ph.D. in chemical engineering from the University of Illinois. Dr. Ayen, now retired, was director of technology for Waste Management, Inc. He has extensive experience in the evaluation and development of new technologies for the treatment of hazardous, radioactive, industrial, and municipal waste. Dr. Ayen managed all aspects of Waste Management's Clemson Technical Center, including treatability studies and technology demonstrations for the treatment of hazardous and radioactive waste. His experience includes 20 years at Stauffer Chemical Company, where he was manager of the Process Development Department at Stauffer's Eastern Research Center. Dr. Ayen has published extensively in his fields of interest.

Robert A. Beaudet is chair of the NRC Committee on Review and Evaluation of Alternative Technologies for Demilitarization of Assembled Chemical Weapons (I and II). He received his Ph.D. in physical chemistry from Harvard University and has served on U.S. Department of Defense committees to address offensive and defensive chemical warfare. Dr. Beaudet was chair of an Army Science Board committee that addressed chemical detection and trace gas analysis and for two years was chair of an Air Force technical conference on chemical warfare decontamination and protection. He has served on the NRC Committee on Chemical and Biological Sensor Technologies and the Committee on Energetic Materials and Science Technology. Most of his career has been devoted to research on molecular structure and molecular spectroscopy. Dr. Beaudet was a member of the Board on Army Science and Technology and served as the BAST liaison to the Review and Evaluation of the Army Chemical Stockpile Disposal Program Committee during the development of nine reports. He is the author or coauthor of more than 100 technical reports and papers.

Lisa M. Bendixen is a principal in the environment and risk practice at Arthur D. Little, Inc. Since joining the company in 1980, Ms. Bendixen has been involved in risk management and risk assessment studies for numerous industries. She is the secretary of the NRC Transportation Research Board's Committee on Hazardous Materials and was the U.S. delegate to the International Electrotechnical Commission's working group on risk analysis until early 1999. She was a member of the NRC Committee on Fiber Drum Packaging for Transporting Hazardous Materials and is past chair of the Safety Engineering and Risk Analysis Division of the American Society of Mechanical Engineers (ASME). She has been involved in many studies on the chemical demilitarization of M55 rockets, including the identification and quantification of failure modes leading to agent release based on a generic disposal facility design; evaluations of sources of risk in separating agent from energetic components in the rocket; and preparation of criteria for evaluating storage, transportation, and on-site disposal options. Ms. Bendixen earned an M.S. in operations research at the Massachusetts Institute of Technology.

Joan B. Berkowitz, managing director of Farkas Berkowitz and Company, has extensive experience in environmental and hazardous waste management, technologies for the cleanup of contaminated soils and groundwater, and a strong background in physical and electrochemistry. She has contributed to several Environmental Protection Agency (EPA) studies, has been a consultant on remediation techniques, and has assessed various destruction technologies. Dr. Berkowitz is the author of numerous publications on hazardous waste treatment and environmental subjects. She was a member of the NRC panel on Review and Evaluation of Alternative Chemical Disposal Technologies and is currently a member of the NRC Committee on Review and Evaluation of Alternative Technologies for Demilitarization of Assembled Chemical Weapons (I and II). She has a Ph.D. from the University of Illinois in physical chemistry.

Judith A. Bradbury, technical manager at Battelle Pacific Northwest National Laboratory, is currently evaluating public involvement programs across the U.S. Department of Energy (DOE) complex. She previously participated in a series of evaluations of the effectiveness of DOE's 12 site-specific advisory boards and led an assessment of community concerns about incineration and perspectives on the U.S. Army Chemical Weapons Disposal Program. Dr. Bradbury is a member of the Risk Assessment and Policy Association. She earned a B.S. in sociology from the London School of Economics, an M.A. in public affairs from Indiana University of Pennsylvania, and a Ph.D. in public and international affairs from the University of Pittsburgh.

Martin C. Edelson has been a member of the staff at the NASA Ames Laboratory since 1977 and is an adjunct associate professor of mechanical engineering at Iowa State University. His research interests include risk communication and the development of laser-based methods for materials processing and characterization. Dr. Edelson was a member of the Munitions Working Group and the DOE Laboratory Directors Environmental and Occupational/Public Health Standards Steering Group. He currently represents the Ames Laboratory on the DOE Strategic Laboratory Council and the Subsurface Contamination Focus Area Lead Laboratory. Dr. Edelson is a technical editor of *Risk Excellence Notes*, a publication funded by the DOE Center for Risk Excellence. He earned a B.S. in chemistry and an M.A. in physical chemistry from City College of New York and a Ph.D. in physical chemistry from the University of Oregon.

Sidney J. Green (NAE) is chairman and chief executive officer of TerraTek, a geotechnical research and services firm in Salt Lake City focused on natural resource recovery, civil engineering, and defense problems. Previously, he worked at General Motors and the Westinghouse Research Laboratory. He has an extensive background in mechanical engineering, applied mechanics, materials science, and geoscience applications and is a former member of the NRC Geotechnical Research Board. He was named Outstanding Professional Engineer of Utah and is the recipient of the ASME Gold Medallion Award and the Lazan Award from the Society of Experimental Mechanics. Mr. Green received a B.S. from the University of Missouri-Rolla and an M.S. from the University of Pittsburgh, both in mechanical engineering, and a Degree of Engineering from Stanford University.

Paul F. Kavanaugh, an engineering management consultant, was the director of government programs for Rust International, Inc., and director of strategic planning for Waste Management Environmental Services. During his military service, he served with the U.S. Army Corps of Engineers, DOE, and the Defense Nuclear Agency and managed engineering projects supporting chemical demilitarization at Johnston Atoll. He earned a B.S. in civil engineering from Norwich University and an M.S. in civil engineering from Oklahoma State University. Brigadier General Kavanaugh is a fellow of the Society of American Military Engineers.

Douglas M. Medville recently retired from MITRE as program leader for chemical materiel disposal and remediation. He has led many analyses of risk, process engineering, transportation, and alternative disposal technologies and has briefed the public and senior military officials on the results. Mr. Medville led the evaluation of the operational performance of the Army's chemical weapon disposal facility on Johnson Atoll and directed an assessment of the risks, public perceptions, environmental aspects, and logistics of transporting recovered nonstockpile chemical warfare materiel to candidate storage and disposal destinations. Previously, he worked at Franklin Institute Research Laboratories and General Electric. Mr. Medville earned a B.S. in industrial engineering and an M.S. in operations research, both from New York University.

Winifred G. Palmer is a toxicologist with the Henry M. Jackson Foundation for the Advancement of Military Medicine, where she has been working under a five-year grant from the U.S. Army Center for Health Promotion and Preventive Medicine. Between 1989 and 1996, she was a toxicologist for the U.S. Army at Fort Detrick, Maryland. Her recent work has included assessments of health risks associated with chemical warfare agents, the development of a military field water-quality standard for the nerve agent BZ, the development of the military air-quality standard for fog oil, and studies on the bioavailability of TNT and related compounds in composts of TNT-contaminated soils. Dr. Palmer is a member of the Society of Toxicology, and her numerous publications span more than two decades of work in the field. She has a B.S. in chemistry and biology from Brooklyn College and a Ph.D. in biochemistry from the University of Connecticut.

James P. Pastorick is president of GEOPHEX UXO, Ltd., an unexploded ordnance (UXO) remediation firm based in Alexandria, Virginia, that specializes in UXO planning and management consulting and the implementation of advanced geophysical UXO detection methods. Since he retired from the U.S. Navy as an explosives ordnance disposal officer and diver in 1989, he has been working on civilian UXO clearance projects. Prior to starting his present company, he was the senior project manager for UXO projects at UXB International, Inc., and the IT Group.

William J. Walsh is an attorney and partner in the Washington, D.C., office of Pepper Hamilton LLP. Prior to joining Pepper, he was a section chief in the EPA Office of Enforcement. His legal experience encompasses environmental litigation on a broad spectrum of issues pursuant to a variety of environmental statutes, including the Resources Conservation and Recovery Act, the Toxic Substances Control Act, and personal injury litigation. He represents trade associations, including the Biotechnology Industry Organization, in rule-making and other public policy advocacy; represents individual companies in environmental actions (particularly in negotiating innovative solutions to environmental disputes); and advises technology developers and users on taking advantage of the incentives for, and eliminating the regulatory barriers to, the use of innovative environmental technologies. He previously served on the NRC Committee on Groundwater Cleanup Alternatives and the Committee on the Use of Groundwater Models in the Regulatory Programs. Currently, he is a member of the NRC Committee on Environmental Remediation at Naval Facilities. Mr. Walsh holds a J.D. from George Washington University, and a B.S. in physics from Manhattan College.

Ronald L. Woodfin is a recently retired staff member of Sandia National Laboratories, where he coordinated work on mine countermeasures and demining, including sensor development. He is currently an adjunct professor of mathematics at Wayland Baptist University, Albuquerque Campus. Previously, he worked at the Naval Weapons Center, Naval Undersea Center, and Boeing Commercial Airplane Division. Dr. Woodfin has been an invited participant at several international conferences on demining and has served on an advisory task force on humanitarian demining for the General Board of Global Ministries of the United Methodist Church. He also serves as pastor of Cedar Crest Baptist Church, Cedar Crest, New Mexico. Dr. Woodfin earned a B.S. in engineering from the University of Texas and an M.S. in aeronautics and astronautics and a Ph.D. in engineering mechanics from the University of Washington.

B

Committee Meetings and Other Activities[1]

First Committee Meeting, June 15–17, 1999
National Research Council, Washington, D.C.

Presentations:

Opening Remarks, Program Update, Year 2 Task and Sponsor Expectations
Wayne Jennings
Project Manager, Non-Stockpile Chemical Materiel (PMNSCM)

Current Status of Rapid Response System (RRS)
Larry Friedman
PMNSCM

Current Status of Munitions Management Device (MMD)-1/2/3
Alan Caplan and Jerry Hawks
PMNSCM

Current Status of the Explosive Destruction System (EDS) and Single CAIS Neutralization System (SCANS)
Mike Duggan and Ed Doyle
PMNSCM

Current Status of Munitions Assessment and Processing System (MAPS), Portable Isotopic Neutron Spectroscopy (PINS), Secondary Ion Mass Spectrometry (SIMS), and Raman Spectroscopy
Ed Doyle and Bill Brankowitz
PMNSCM

RRS and MMD Neutralization Wastes and Existing/Expected Recovered Chemical Warfare Materiel to be Treated and Air Monitoring
John Gieseking
PMNSCM

Plan and Current Status of the Army's Assembled Chemical Weapons Assessment (ACWA) Program
Carl Eissner
Soldier and Biological Chemical Command

Plan and Current Status of the Alternative Technologies and Approaches Project
Nick Levitt
Project Manager, Alternative Technologies and Approaches

Plan and Current Status of Mitretek Technology Survey Project for PMNSCM
George Bizzigotti
Mitretek

Public and Stakeholder Concerns
Elizabeth Crowe
Non-Stockpile Citizens Coalition

Plan and Current Status of Stone & Webster Technology Survey Project for PMNSCM
Joseph Cardito
Stone & Webster Engineering Corporation

---

[1] The committee gathered additional information via telephone conference calls and by other means. Details are available on line at: *http://www4nas.edu/cets/dmst.nsf/*

APPENDIX B

Closing Comments
Col. Ned Libby
PMNSCM

Second Committee Meeting
Tour of Training Facility and Briefings
August 31–September 1, 1999
Aberdeen Proving Ground, Maryland

Presentations:

Tour of Chemical Demilitarization Training Facility
Andrew Roach
Project Manager, Chemical Demilitarization Operations

Tour of Explosive Destruction System
Ray DiBerardio
Non-Stockpile Chemical Materiel

ACWA Report Findings
Robert Beaudet, chair
Committee on Review and Evaluation of Alternative Technologies for Demilitarization of Assembled Chemical Weapons

Non-Stockpile Program Status
LTC Chris Ross
Program Manager, NSPCM

Third Committee Meeting, October 14–15, 1999
National Research Council, Woods Hole, Massachusetts

Presentations:

Non-Stockpile Program Status
LTC Chris Ross/Wayne Jennings
NSPCM

ACWA Program
James Richmond
ACWA Program

Stone & Webster Progress Report
Joseph Cardito
Stone & Webster

Mobile Alternative Demilitarization Technologies
Dr. Francis W. (Bill) Holm
Consultant

Fourth Committee Meeting, December 15–16, 1999
National Research Council, Washington, D.C.

Two-day writing session. No presentations.

Fifth Committee Meeting, February 22–23, 2000
National Research Council, Washington, D.C.

Presentations:

Stone & Webster Progress Report
Joseph Cardito
Stone & Webster

Non-Stockpile Chemical Weapons Citizens Coalition
Elizabeth Crowe
Non-Stockpile Chemical Weapons Citizens Coalition

Sixth Committee Meeting, May 8–9, 2000
National Research Council, Washington, D.C.

Presentation:

U.S. Army Non-Stockpile Chemical Materiel Product (NSCMP) Project Overview/Status
William R. Brankowitz
Deputy Product Manager, NSCMP

Seventh Committee Meeting, August 29–30, 2000
National Research Council, Woods Hole, Massachusetts

Presentations:

Non-Stockpile Program Status
LTC Chris Ross/Wayne Jennings
NSPCM

Stone & Webster Progress Report
Joseph Cardito
Stone & Webster

Transportable Batch Hydrothermal Oxidizer for Non-stockpile Chemical Material
Brent L. Haroldson and Ben Wu
Sandia National Laboratories

SITE VISITS

Dugway Proving Ground, Utah, and Deseret Chemical Depot, Utah, August 3–4, 1999

Site Team

John C. Allen
Joan B. Berkowitz
Judith A. Bradbury
Martin C. Edelson
Sidney J. Green
Douglas M. Medville
Winifred G. Palmer
Ronald L. Woodfin
NRC Staff
Michael Clarke
Delphine D. Glaze
Gregory Eyring

Tour of Rapid Response System at Deseret Chemical Depot
Hosts: Michael Nuttle, Harold Oliver, Walter Levi, Brett Simms

Meeting with Utah Citizens Advisory Council (CAC)
CAC members present: Dave Ostler, Rosemary Holt, John Matthews, Dan Bauer

Tour of Supercritical Water Oxidation Facility at Dugway Proving Ground
Hosts: William Dement, Charles Donaldson, Andrew Nifsi, Beryl Schwartz, Robert Edgin, Donald Spina, Bud Salzburg, Michael Spritzer

ViVendor Test Facility, San Diego, California, March 2, 2000

Site Team

Robert A. Beaudet
Judith A. Bradbury
Martin C. Edelson
Douglas M. Medville
Greg Eyring
Jane Williams (Sierra Club)

Tour of Ventless Incineration at ViVendor Test Facility
Host: Gere Johansing

CerOx Corporation, Reno, Nevada, March 21, 2000

Site Team

Richard J. Ayen
Robert A. Beaudet
Joan B. Berkowitz
Paul F. Kavanaugh
Sterling J. Rideout (NRC study director)

Tour of CerOx Corporation Electrochemical Cerium Process
Hosts: Dr. Steven Oberg (University of Nevada-Reno), Marty Scanlon, Thomas Neustedter, Fred Coppotelli

Meeting with EPA Regulators, Stephen Heare, Jeffrey Gaines, and Carl Duly (via telephone), Rosslyn, Virginia, April 28, 2000

Site Team

Judith A. Bradbury
Douglas M. Medville
William J. Walsh
Sterling J. Rideout (NRC study director)